AF555939

Energy, Environment, Ecology and Society

Energy, Environment, Ecology and Society

Anil Kumar

and

K. Sudhakar

Assistant Professors, Department of Energy,
Maulana Azad National Institute of Technology,
Bhopal - 462051, MP.

An unit of **BSP Books Pvt., Ltd.**

4-4-309/316, Giriraj Lane, Sultan Bazar,
Hyderabad - 500 095
Phone : 040 - 23445605, 23445688

Published by :

BS Publications
An unit of **BSP Books Pvt., Ltd.**

4-4-309/316, Giriraj Lane, Sultan Bazar,
Hyderabad - 500 095
Phone : 040 - 23445605, 23445688
e-mail : info@bspbooks.net

ISBN : 978-93-85443-15-3 (HB)

Dedicated to

Teachers, Parents and Families

Preface

Population, exuberant growth of urbanization, decline of cultivable lands, growing number of vehicle on the roads, deforestation, industrialization, changing pattern of consumption and exploitation of natural resourses by human activities have all threatened our basic survival on earth. In order to protect our globe from the environmental degradation, it is necessary to know the various factors by all human being. This book is written to provide a clear and authoritative introduction to the subject of Energy, Environment, Ecology and Society.

The subject "Energy, Environment, Ecology and Society" has been included for all branches of Engineering and Technology by RGPV in Madhya Pradesh. In order to know all aspects of the subject, more books and literature have to be referred. It is the daunting task for the students and faculties of all Institutions. To avoid the complexities, this book entitled "Energy Environment Ecology and Society" has been brought out. It follows the sequence of topics, prescribed in the course syllabus.

This book has been divided into eight chapters. Chapter 1 deals with Energy Sources, Chapter 2 deals with Ecosystem and Biodiversity, third chapter deals with Air pollution, Chapter 4, 5 and 6 deal with Sound, Water, and Soil pollution, then Chapter 7 deals with Society and Ecology and Chapter 8 deals with Ethics. Similarly in each chapters review questions are given to help the students for university examination.

- *Authors*

Acknowledgement

It is a pleasure for us to acknowledge the help, advice and encouragement which we have been received from colleagues and students of Department of Energy, MANIT, Bhopal.

The first author would like to thank Prof. G. N. Tiwari, Centre for Energy Studies, Indian Institute of technology, Delhi, Hon. Vice Chancellor Prof. Piyush Trivedi, Prof. V.K. Sethi, Prof. Vice Chancellor and Prof. S. N. Verma, Dean (Industrial Technology), Rajiv Gandhi Proudyogiki Vishwavidyalaya, Bhopal in building up his teaching career during his earlier formative years. He is extremely grateful to his wife, Ms. Abhilasha Kumar and his son Mr. Tijil Kumar for their constant support in all his endeavors.

The second Author expresses his sincere thanks to the following persons

- Dr. M. Premalatha, Associate Professor, CEESAT, NIT Trichy,
- Hon. Chairman Thiru L. Prakashmul Chordia and Ex. Director L. Subramaniam, Annai Teresa College of Engineering, Thirunavalur, TN.
- Prof. R. L. Sawhney and Prof. S. P. Singh, SEES, DAVV, Indore.

We extend our acknowledgement to Mr. Rajesh and Mr. Himanshu for their assistance. Last but not the least, we have to thank the publisher, BS Publications, Hyderabad for their assistance and encouragement.

As authors, we bear responsibility for all interpretation, opinions and errors in this work, however, many have helped us and we express our gratitude to all of them. We would thankfully welcome constructive suggestions and comments for further improvement of the book

- Authors

Contents

Chapter 3

Air Pollution

Chapter 4

Sound Pollution

Chapter 5

Water Pollution

Chapter 8

Ethics

CHAPTER 1

Energy

Energy- Sources of Energy: Renewable & Non Renewable, Fossil fuel, Coal, oil, Gas, Geothermal, Hydrogen, Solar, Wind, Biomass, Hydal, Nuclear sources.

1.1 Energy

Energy is the basis of human life. There is hardly any activity or moment that is independent of energy. Every moment of the day we are using energy. Over the past few decades, energy is the backbone of technology and economic development.

Energy, 'the ability to do work', is essential for meeting basic human needs, extending life expectancy and providing comfort in living standards.

The standard of living of the people of any country is considered to be proportional to the energy consumption by the people of that country. In one sense, the disparity from country to country arises from the extent of accessible energy for the citizens of each country. Unfortunately, the world energy demands are mainly met by the fossil fuels today.

1.2 Types of Energy Resources

Energy can be considered in two categories - primary and secondary. Fig. 1.1 shows Classification of Energy Sources.

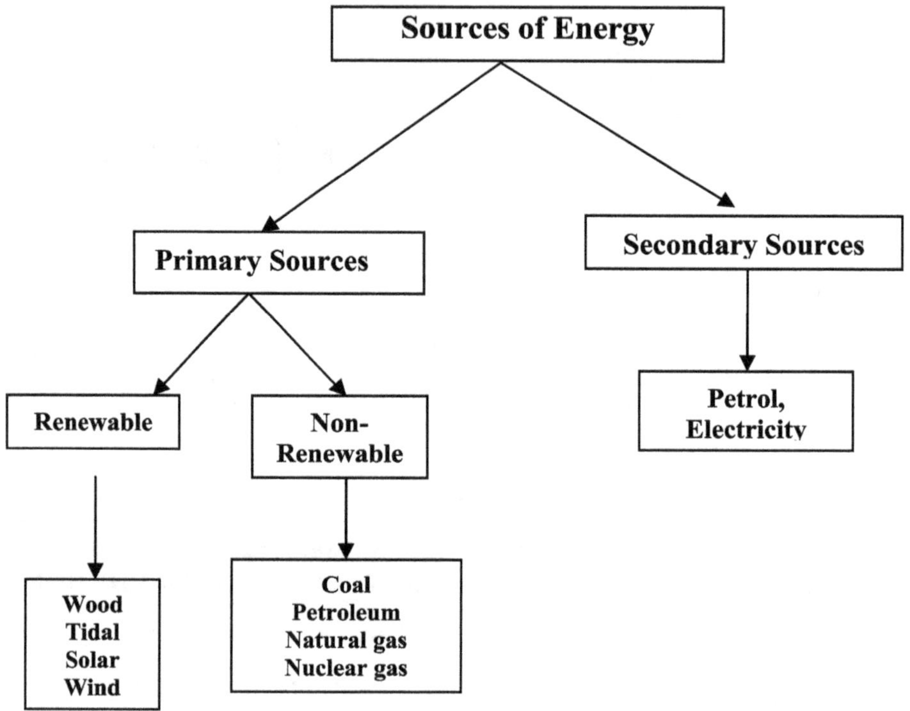

Fig. 1.1 Classification of Energy Sources

(i) **Primary energy** is energy in the form of natural resources, such as wood, coal, oil, natural gas, natural uranium, wind, hydro power, and sunlight. These energy resources are mined or otherwise obtained from the environment.

Examples:

(a) *Fossil fuels:* coal, lignite, crude oil, Natural gas etc.

(b) *Nuclear fuels:* Uranium, Thorium, other nuclear used in friction reaction.

(c) *Hydro energy:* It is energy from falling water, with the help of a turbine.

(d) *Geo thermal:* The heat from the underground stream.

(e) *Solar energy:* Electromagnetic radiation from the Sun.

(f) *Wind energy:* The energy from moving air used by wind mills.

(g) *Tidal energy:* The energy associated with the rise and fall of the tidal wave.

(ii) Secondary energy is the more useable forms to which primary energy may be converted, such as electricity and petrol. Primary energy can be renewable or non-renewable: Renewable energy sources include solar, wind and wave energy, biomass (wood or crops such as sugar), geothermal energy and hydro power. Non-renewable energy sources include the fossil fuels - coal, oil and natural gas, which together provide 80% of our energy today, plus uranium.

(iii) Renewable energy or inexhaustible or non-conventional sources can be defined as energy obtained from the continuous or repetitive current of energy recurring in the natural environment. These energy sources are produced continuously or that are available indefinitely. Example: solar energy, wind energy, hydro, biomass, ocean tide, geothermal, etc. This is also referred as Non-conventional source of energy.

(iv) Non-renewable energy or exhaustible or conventional sources are available in limited amount and that can be exhausted. They deplete over a longer period of time.Hence, they cannot be replenished in the quantities they are being consumed in a given period of time. Examples are fossil fuels of coal, oil and natural gas and nuclear fuels etc. these types of energy also called Conventional energy sources (Fig. 1.2).

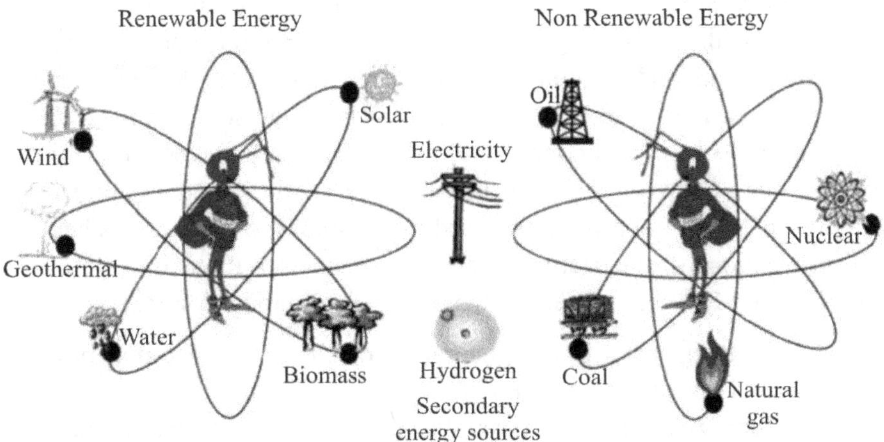

Fig. 1.2 Various forms of renewable and non-renewable energy sources

1.3 Energy Status of India

India's energy status is not promising. Presently, the country consumes about 100 million tones of coal and 32.5 million tones of oil annually. Official estimate report that 40 billion tones of coal are available but only one half this is recoverable which means it is less than the projected demand of 23 billion tones of coal till the year 2020.

India's oil deposit is about 400 million tones as against the world oil reserve of 750,000 million tones. Gas reserves of our country are about 100 million cubic meters, as against world's reserves of 63,000 million cubic meters. Here, one can conclude that the energy Scenario of India is blank.

1.4 Fossils Fuels

Fossils fuels (oil, coal, natural gas) are energy rich substances that have formed from the remains of organisms that lived 200 to 500 million years ago. During the stage of the Earth's evolution, large amount of dead organic matter had collected. Over million of years, this matter was buried under layers of sediment and converted by heat and pressure into coal, oil and natural gas.

Chemically, fossil fuels largely consist of hydrocarbons, which are compounds of hydrogen and carbon. Some fossils fuel also contains smaller quantities of other compounds. After the accumulating sediments exerted increasing heat and pressure for millions of years on the ancient organisms hydrocarbons were formed. Most common among them are petroleum, coal and natural gas. However Geologists have identified other types of hydrocarbon rich deposits, which can serve as fuels. Such deposits are: oil shale, tar sands and gas hydrates. However, they are not widely used due to the fact that they are very costly to extract and refine. Majority of fossil fuels are being used in transportation, industries heating and generation of electricity. Crude petroleum is refined into gasoline; diesel and jet fuel that power the world's transportation system.

Coal is mostly used in the generation of electricity (thermal power). Natural gas is used for commercial and domestic purposes like heating, air conditioning and as fuels for stoves and for other heating appliances. Once we discovered the fossil fuel we began consuming them at an increasing rate. From 1859 to 1969, total oil production was227 billion barrels (1 barrel=159 lts). 50% of this total was extracted during the first 100 years, while the next 50% was extracted in next 10 years. Today, fossil fuels are considered to be non-renewable for the reason that their consumption rate is far in excess of the rate of their formation.

1.5 Coal

Coal is the most abundant fossil fuel on the planet, with current estimates from 216 years global recoverable reserves to over 500 years at current usage rates. About 250 to 350 million years ago coal was formed on earth in hot, damped regions. But the global distribution of coal is non-uniform like any other mineral deposits or for that matter petroleum. For instance one half of the world's known reserves of coal are in the United States of America. Almost 27350 billion metric tones of known coal deposits occur on our planet. Out of which about 56% are located in Russia, 28% in USA and Canada.

Mainly, there are three types of coal:

Anthracite or hard coal (90% carbon content)

Bituminous or soft coal (85% carbon content)

Lignite or brown coal (70% carbon content)

The present annual extraction rate of coal is about 3000 million metric tones, at this rate coal reserves may lasts for about 200 hundred years and if its use is increased by 2% per year then it will last for another 65 years.

(i) ***Transformation of peat to coal:*** Coal is formed by the partial decomposition of vegetable matter and is primarily organic in nature. It is well studied as a sedimentary rock. Coal is a complex organic natural product that has evolved from precursor materials over millions of years. It is believed that the formation of coal occurred over geological times in the absence of oxygen there by promoting the formation of a highly carbonaceous product through the loss of oxygen and hydrogen from the original precursor molecules. Simplified representation of coal maturation by inspection of elemental composition is presented in Fig. 1.3.

		Composition, wt%			H/C	
		C	H	O		
Increasing	Wood	49	7	55	1.7	Increasing
pressure,	Peat	60	6	34	1.2	aromatization
temperature,	Lignite coal	70	5	25	0.9	loss of
time	Sub-bituminous coal	75	5	20	0.8	oxygen
↓	Bituminous coal	85	5	10	0.7	↓
	Anthracite coal	94	3	3	0.4	

Fig. 1.3 Maturation of coal

Each class implies higher carbon content than the preceding one, e.g., bituminous coals have greater carbon content than sub-bituminous coals. As shown coals are composed of C, H, O, N and S. A progressive change in composition is found through the coal rank series.

A coal of a certain level of maturity, or degree of metamorphosis from the peat, is said to be of certain rank. The different types of coals which are clearly recognizable by their different properties and appearance can be arranged in the order of their increasing metamorphosis from the original peat material. They are:

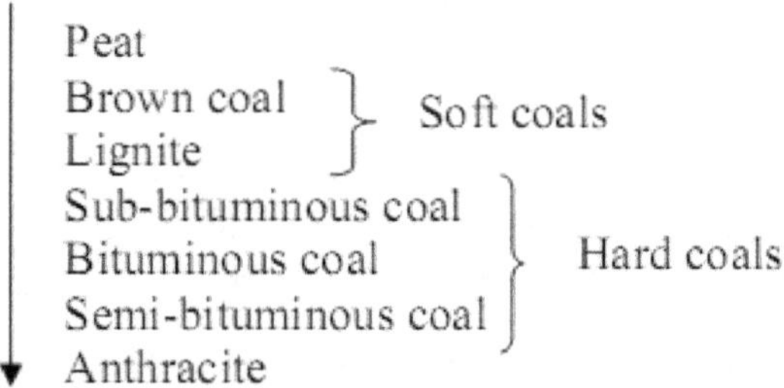

The most highly changed material, this is the final member of the series of coals formed from peat is Anthracite. Each member of the series represents a greater degree of maturity than the preceding one. The whole is known as the "*peat-to-anthracite series*".

(ii) ***Coal for the generation of electricity*:** From coal to electricity generation is not a single step conversion. The purpose of coal is only to get heat energy. This heat energy in turn is used to convert water to steam. The steam makes the propeller-like blades of the turbine to rotate at high speeds. A generator connected to the turbine converts mechanical energy to electrical energy. The various components and steps involved in the generation of electricity are depicted in Fig. 1.4.

The electricity generated is transformed into the higher voltages upto 4, 00, 000 volts and used for economic, efficient transmission via power line grids. When it nears the point of consumption, such as our homes, the electricity is transformed down to the safer 100-250 V used in the domestic market.

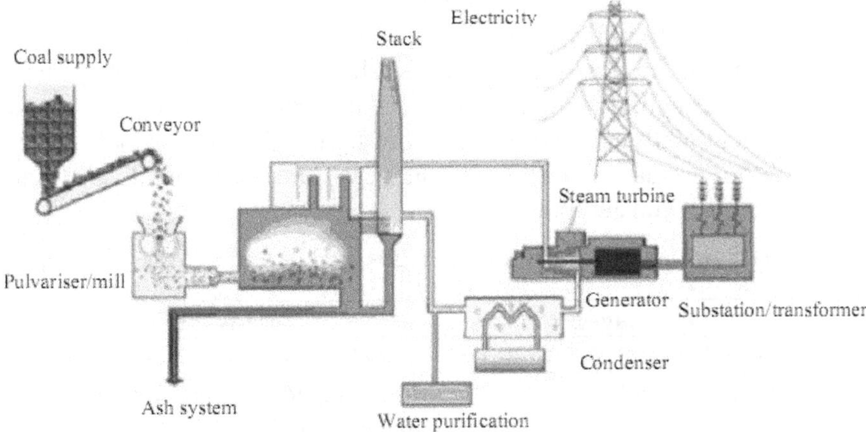

Fig. 1.4 Schematic representation of the processes taking place in a power plant

Modern pulverized coal combustion technology is well- developed and accounts for over 90% of coal fired capacity world wide. Improvements are being made in the direction of producing more electricity from less coal being used i.e., to improve the thermal efficiency of the power station.

(iii) ***Coal Reserves – From Indian perspective*:** Coal is the predominant energy source (58%) in India, followed by oil (27%), natural gas (7%), lignite (4%), hydropower (3%) and nuclear power (0.22%). India has about 5% of world's coal reserve and that too not of vary good quality in term of heat capacity.

Major coal fields in India are found in Jharkand, Bihar, West Bengal, Madhya Pradesh, Maharastra, Assam, Andhra Pradesh, Orissa, Tamil Nadu and Kashmir. Jharia (Jharkand) and Raniganj (West Bengal) are the biggest and best coalfields of the country. Unlike the coals in Europe and America, Indian coals have high percentage of mineral matter, most of which is finely disseminated and intimately mixed with the coal substance.

Nearly 32.98 % of coal deposits of India are in Jharkand. Its mines are in Jharia, Chandrapure, Bokaro, Ramgarh, Kamapur, Charhi and also in Rajmahal and Daltonganj area. Raniganj coal field is the largest coalfield in India, belonging to the Gondwana Super group (Gondwana is a geological term which refers to a certain rock system which is about 200 million years old. Most of the Indian coals belong to this group

Lignite is found mainly at Neyveli in Tamil Nadu. Minor coal fields exist in Andhra Pradesh, Kashmir and Assam. Assam coals have very high sulphur content (3-8 %). Kashmir coals are artificial anthracite converted from lignite deposits. Coal deposits of the Tertiary era (60 million years old) are found in Assam, Rajasthan and Jammu.

(iv) Coal: Advantages & Disadvantages coal given below

Advantages	Disadvantages
Ample supplies (225 Years)	Very high environmental impact
High net energy yield	Several land disturbance air pollution and water pollution
Low cost (with huge substitutes)	High land use (including mining)
Mining and combustion technology well developed	Severe threat to human health
Air pollution can be reduced with developed technology	High carbon dioxide emissions when burnt Releases radio active particles and mercury into air.

1.6 Crude Oil / Petroleum

Crude oil is oily, flammable, thick dark brown or greenish liquid that occurs naturally in deposits, usually beneath the surface of the earth; it is also called as Petroleum. Petroleum means rock oil, (Petra – rock, elaion – oil, Greek and oleum – oil, Latin), the name inherited for its discovery from the sedimentary rocks. It is used mostly for producing fuel oil, which is the primary energy source today. Convenience of petroleum or mineral oil and its greater energy content as compared to coal on weight basis has made it the lifeline of global economy.

Petroleum is cleaner fuel when compared to wood or coal as it burns completely and leaves no residue. Petroleum is often considered the lifeblood of nearly all other industry. For its high energy content (Table 1.1) and ease of use, petroleum remains as the primary energy source.

Table 1.1 Energy density of different fossil fuels

Fuel	Energy Density
Petroleum or Crude oil	45 MJ/Kg
Coal	24 MJ/Kg
Natural Gas	34 – 38 MJ/m^3

Petroleum is unevenly distributed like any other mineral. There are 13 countries in the world having 67% of the petroleum reserves which together form the OPEC (Organization of petroleum exporting countries). Six regions in the world are rich in petroleum – USA, Mexico, Russia and West Asian countries. Saudi Arabia oil producing has one fourth of the world oil reserves. The total oil reserves of our planet is about 356.2 billion metric tones out of this annually we are exporting about 28% million metric tones. Hence the existing reserves would last for about 40 – 50 years. About 40% of the total energy consumed in the entire world is now contributed by oil.

The oil bearing potential of India is estimated to be above one million square kilometers is about one third of the total geographic area. Northern plains in the Ganga-Brahmaputra valley, the coastal strips together with their off-shore continental shelf (Bobay Hihgh), the plains of Gujarat, the Thar Desert and the area around Andaman and Nicobar Islands.

(i) **Composition of Petroleum:** Petroleum is a combination of gaseous, liquid and solid mixtures of many alkanes. It consists principally of a mixture of hydrocarbons, with traces of various nitrogenous and sulfurous compounds. Gaseous petroleum consists of lighter hydrocarbons with abundant methane content and is termed as 'natural gas'. Liquid petroleum not only consists of liquid hydrocarbons but also includes dissolved gases, waxes (solid hydrocarbons) and bituminous material.

Solid petroleum consists of heavier hydrocarbons and this bituminous material is usually referred to as bitumen or asphalt. Along with these, petroleum also contains smaller amounts of nickel, vanadium and other elements (≈ 1000 ppm). The elemental composition of petroleum varies greatly from crude oil to crude oil. Table 2 shows overall composition of petroleum.

Table 1.2 Overall Composition of Petroleum

Element	Percentage composition
Carbon	83.0-87.0
Hydrogen	10.0-14.0
Nitrogen	0.1-2.0
Sulphur	0.05-6.0
Oxygen	0.05-1.5

(ii) ***Petroleum Refining*:** Raw oil or unprocessed crude oil is not very useful in the form it comes in out of the ground. It needs to be broken down into parts and refined before use in a solid material such as plastics and foams, or as petroleum fossil fuels as in the case of automobile and air plane engines. An oil refinery is an industrial process plant where crude oil is processed in three ways in order to be useful petroleum products (i) Separation - crude oil separates into various fractions and oil can be used in so many various ways because it contains hydrocarbons of varying molecular masses and lengths such as paraffins, aromatics, naphthenes (or cycloalkanes), alkenes, dienes, and alkynes.

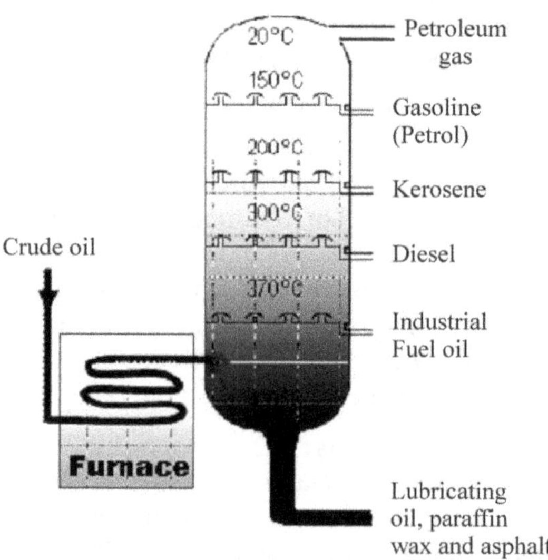

Fig. 1.5 Schematic of the distillation of crude oil

Hydrocarbons are molecules of varying length and complexity made of hydrogen and carbon. The trick in the separation of different streams in oil refinement process is the difference in boiling points between the hydrocarbons, which means they can be separated by distillation. Fig. 1.5 shows the typical distillation scheme of an oil refinery. (ii) Conversion –Once separated and any contaminants and impurities have been removed, the oil can be either sold with out any further processing, or smaller molecules such as isobutene and propylene or butylenes can be recombined to meet specified octane number requirements by processes such as alkylation or less commonly, dimerization. (iii) Finishing- In this process, purification of the product streams is done.

(iii) Products of oil refinery:

Asphalt: Asphalt is a sticky, black and highly viscous liquid or semi-solid that is present in most crude petroleum. Asphalt is composed almost entirely of bitument. Asphalt is rather hard to transport in bulk so it is sometimes mixed with diesel oil or kerosene before shipping. Upon delivery, these lighter materials are separated out of the mixture. The largest use of asphalt is for making asphalt concrete for pavements. Roofing shingles account for most of the remaining asphalt consumption. Other uses include cattle sprays, fence post treatments, and waterproofing for fabrics. The ancient middle-east natural asphalt deposits were used for mortar between bricks and stones, ship caulk, and waterproofing.

Diesel Fuel: Petroleum derived diesel is composed of about 75% saturated hydrocarbons (primarily paraffins including n, iso, and cycloparaffins), and 25% aromatic hydrocarbons (including naphthalens and alkylbenzenes). The average chemical formula for common diesel fuel is $C_{12}H_{26}$, ranging from approximately, $C_{10}H_{22}$ to $C_{15}H_{32}$. It is obtained in the fractional distillation of crude oil between 250 °C and 350 °C at atmospheric pressure. Diesel is considered to be a fuel oil and is about 18% denser than gasoline. The density of diesel is about 850 grams per liter whereas gasoline has a density of about 720 g/l, or about 18% less. Diesel is generally simpler to refine than gasoline and often costs less. Diesel contains approximately 18% more energy per unit of volume than gasoline, which, along with the greater efficiency of diesel engines, contributes to fuel economy.

Fuel Oil: Fuel oil is a fraction obtained from petroleum distillation, either as a distillate or a residue. Broadly speaking, fuel oil is any liquid petroleum product that is burned in a furnace for the generation of heat. Fuel oil is made of long hydrocarbon chains, particularly alkanes, cycloalkanes and aromatics. The fuel oil is the heaviest commercial fuel that can be obtained from crude oil, heavier than gasoline and naphtha. The boiling point ranges from 175 to 600 °C, and carbon chain length, 20 to 70 atoms. These are mainly used in ships with varying blending proportions.

Gasoline: Gasoline (or petrol) is a petroleum-derived liquid mixture consisting primarily of hydrocarbons, used as fuel in internal combustion engines. Gasoline is separated from crude oil via distillation, called natural gasoline, will not meet the required specifications for modern engines (in particular octane rating), but these streams will form of the blend. The bulk of a typical gasoline consists of hydrocarbons between 5 to 12 carbon atoms per molecule. Overall a typical gasoline is predominantly a mixture of paraffins (alkanes), naphthenes (cycloalkanes), aromatics and olefins (alkenes).

Kerosene: Kerosene is a colourless flammable hydrocarbon liquid. Kerosene is obtained from the fractional distillation of petroleum at 150 °C and 275 °C (carbon chains from C_{12} to C_{15} range).At one time it was widely used in kerosene lamps but it is now mainly used in aviation fuel for jet engines. A form of kerosene known as RP-1 is burnt with liquid oxygen as rocket fuel. Thick black smoke created during combustion due to the low temperature of combustion.

Liquefied petroleum gas: LPG is manufactured during the refining of crude oil, or extracted from oil or gas streams as they emerge from the ground. Liquefied petroleum gas (LPG) is a mixture of hydrocarbon gases used as a fuel in cooking, heating appliances, vehicles, and increasingly replacing fluorocarbons as an aerosol propellant and a refrigerant to reduce damage to the ozone layer. Varieties of LPG bought and sold include mixes that are primarily propane, mixes that are primarily butane, and mixes including both propane and butane, depending on the season. Propylene and butylenes are usually also present in small concentrations. A powerful odorant, ethanethiol, is added so that leaks can be detected easily.

Lubricant: A lubricant is introduced between two moving surfaces to reduce the friction and wear between them. A lubricant provides a protective film which allows for two touching surfaces to be separated, thus lessening the friction between them. Typically lubricants contain 90% base oil (most often petroleum fractions, called mineral oils) and less than 10% additives.

Paraffin: Paraffin is a common name for a group of high molecular weight alkane hydrocarbons with the general formula C_nH2_{n+2}, where n is greater than about 20. It is also called as paraffin wax. It is mostly found as a white, odourless, tasteless, waxy solid, with a typical melting point between about 47 °C to 65 °C. It is insoluble in water, but soluble in ether, benzene, and certain esters. Paraffin is unaffected by most common chemical reagents, but burns readily. Paraffin is used in Candle making, , As anticaking, moisture repellent and dust binding coatings for fertilizers, Solid propellant for hybrid rockets, Sealing jars, cans, and bottles, In dermatology, as an emollient (moisturizer) etc.

Mineral Oil: Mineral oil is a by-product in the distillation of petroleum to produce gasoline. It is chemically-inert transparent colourless oil co mposed mainly of alkanes and cyclic paraffins, related to white petroleum. It is used as transformer oil, to store and transport alkali metals, an ingredient in baby lotions, cold creams, ointments and other pharmaceuticals and cosmetics, Lubrication, Coolant etc.

Tar: Tar is viscous black liquid derived from the destructive distillation of organic matter. Most tar is produced from coal as a byproduct of coke production, but it can also be produced from petroleum, peat or wood. Surprisingly petroleum tar is the most effective, is used in treatment of psoriasis. Tar is a disinfectant substance, and is used as such.

Bitumen: Bitumen is a category of organic liquids that are highly viscous, black, sticky and wholly soluble in carbon disulfide. Asphalt and tar are the most common forms of bitumen. Bitumen in the form of asphalt is obtained by fractional distillation of crude oil. Bitumen being the heaviest and being the fraction with the highest boiling point; it appears as the bottom most fractions. Bitumen in the form of tar is obtained by the destructive distillation of organic matter, usually bituminous coal. Bitumen is primarily used for paving roads. In the past, bitumen was used to waterproof boats, and even as a coating for buildings.

Pitch (resin): Pitch is the name for any of a number of highly viscous liquids which appear solid. Pitch can be made from petroleum products or plants. Petroleum-derived pitch is also called bitumen. Pitch produced from plants is also known as resin or rosin. Pitch flows at room temperature, but extremely slowly. Pitch has a viscosity approximately 100 billion (1011) times that of water.

(iv) Conventional oil: Advantages & disadvantages of conventional oil is given below

Advantages	Disadvantages
• Amply supply for 40-90years	• Need to find substitute within 50 years
• Low cost (with huge substitute)	• Artificially low price
• High net energy yield	• Encourages waste and discourages search for alternative
• Easily transported within and between countries	• Air pollution when burnt
• Low land use	• Released carbon dioxide when burnt
• Technology is well developed	• Moderate water pollution
• Efficient distribution system	

1.7 Natural Gas

Natural gas has emerged as promising fuel due to its environment friendly nature, efficiency, and cost effectiveness. Economically natural gas is more efficient since only 10 % of the produced gas wasted before consumption and it does not need to be generated from other fuels. Natural gas mainly consists of Methane (CH4) along with other inflammable gases like Ethane and propane. Natural gas is least polluting due to its low Sulphur content and hence is clearest source of energy. It is used both for domestic and industrial purposes. Natural gas is used as a fuel in thermal plants for generating electricity as a source of hydrogen gas in fertilizing industry and as a source of carbon in tyre industry.The total natural gas reserves of the world is about 600 000 billion meters, out of this Russia has 34%, Middle East 18%, North America 17%, Africa and Europe 9% each and Asia 6%. Annual production of natural gas is about 1250 billion cubic meters and hence it is expected to last for about 50-100 years. In India gas reserves are found in Tripura, Jaisalmer, off shore areas of Bombay and Krishna-Godavari Delta. Natural gas was formed from the remains of tiny sea animals and plants that died 200-400 million years ago.

(i) ***Physical properties of Natural gas*:** Natural gas is a mixture of light hydrocarbons including methane, ethane, propane, butanes and pentanes. Other compounds found in natural gas include CO_2, helium, hydrogen sulphide and nitrogen. The composition of natural gas is never constant, however, the primary component of natural gas is methane (typically, at least 90%). Methane is highly flammable, burns easily and almost completely. It emits very little air pollution. Natural gas is neither corrosive nor toxic, its ignition temperature is high, and it has a narrow flammability range, making it an inherently safe fossil fuel compared to other fuel sources. In addition, because of its specific gravity (0.60), lower than that of air (1.00), natural gas rises if escaping, thus dissipating from the site of any leak. Natural gas and/or its constituent hydrocarbons are marketed in the form of different products, such as lean natural gas, wet natural gas (liquefied natural gas (LPG)), compressed natural gas (CNG), natural gas liquids (NFL), liquefied petroleum gas (LPG), natural gasoline, natural gas condensate, ethane, propane, ethane-propane fraction and butanes.

(ii) ***Natural Gas production in India*:** Over the last decade, natural gas energy sector gained more importance in India. In 1947 production of natural gas was almost negligible, however at present the production level is of about 87 million standard cubic meters per day (MMSCMD). Oil & Natural Gas Corporation Ltd. (ONGC), Oil India Limited (OIL) and JVs of Tapti, Panna-Mukta and Ravva are the main producers of Natural gas. Western offshore area is major contributing area to the total production. The other areas are the on-shore fields in Assam, Andhra Pradesh and Gujarat States. Smaller quantities of gas are also produced in Tripura, Tamil Nadu and Rajasthan States. Proven world reserves and resources of natural gas equal 3,200 trillion cubic feet. This equals a 60 year supply at present usage rates.

(iii) ***Natural gas*****:** Advantages & disadvantages of natural gas is given below

Advantages	Disadvantages
• Ample supplies (125 years)	• Non-renewable resources
• High net energy yield	• Releases carbon dioxide when burnt
• Low cost (with huge subsidies)	• Methane (a green house gas) can leak from pipelines
• Less air pollution than other fossil fuels	• Shipped across ocean as highly explosive LNG
• Moderate environmental import	• Sometimes burnt off and wasted at wells because of low prices
• Easily transported by pipelines	• Requires pipelines to transport and store
• Low land use	
• Food fuel for fuel cells and gas turbines	

1.8 Geothermal Energy

Geothermal Energy is heat (thermal) derived from the earth (geo). It is the thermal energy contained in the rock and fluid (that fills the fractures and pores within the rock) in the earth's crust.It is believed that the ultimate source of geothermal energy is radioactive decay occurring deep within the earth .In most areas, this heat reaches the surface in a very diffuse state. However, due to a variety of geological processes, some areas, including substantial portions of many western states, are underlain by relatively shallow geothermal resources.

(i) ***Types and application of Geothermal Energy*****:** These resources can be classified as low temperature (less than 90°C or 194°F), moderate temperature (90°C - 150°C or 194 - 302°F), and high temperature (greater than 150°C or 302°F). The uses to which these resources are applied are also influenced by temperature. The highest temperature resources are generally used only for electric power generation. Uses for low and moderate temperature resources can be divided into two categories: direct use and ground-source heat pumps.

(a) Direct use, implies, involves using the heat in the water directly (without a heat pump or power plant) for such things as heating of buildings, industrial processes, greenhouses, aquaculture (growing of fish) and resorts. Direct use projects generally use resource temperatures between 38°C (100°F) to 149°C (300°F).

(b) **Ground-source heat pumps** use the earth or groundwater as a heat source in winter and a heat sink in summer. Using resource temperatures of 4°C (40°F) to 38°C (100°F), the heat pump, a device which moves heat from one place to another, transfers heat from the soil to the house in winter and from the house to the soil in summer.

(c) **Power Generation.** Geothermal Power plant taps the natural supplies of heat energy that have accumulated inside the earth. Fig. 1.6. Shows schematic of geothermal power plant.

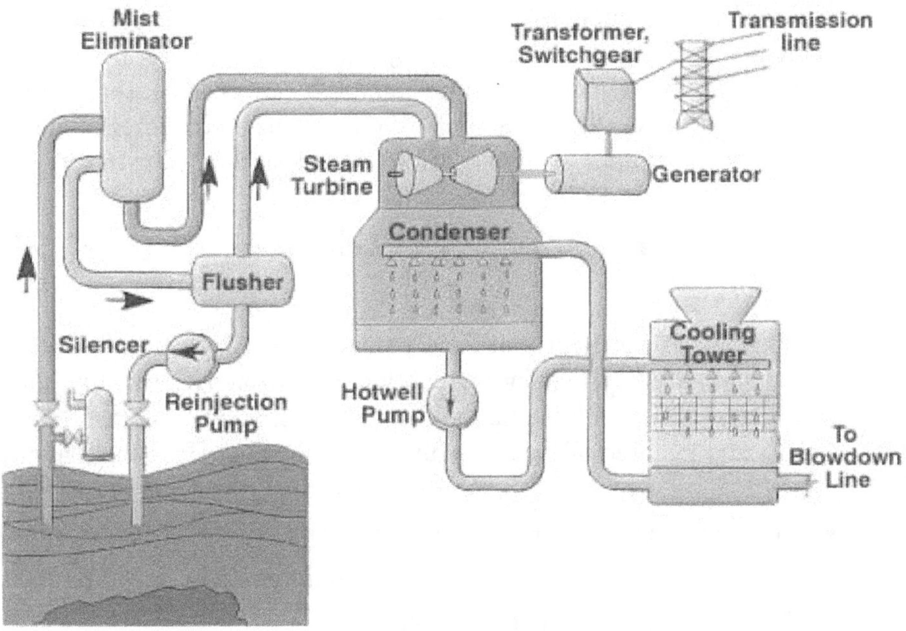

Fig. 1.6 Schematic of Geo-Thermal Power Plant

The geothermal steam turbine is a thermal turbine in which Mother Nature plays the role of a boiler. Since the steam does not contain hot water and its maximum superheating degree is 9°C, no flashing was required. Consequently, direct condensing system utilizing natural steam spurting out from production wells was employed for the power plant. Superheating steam is changed into wet steam from the turbine second stage and is expanded up to 102 mmHg absolute,

52°C. When contacting and mixed with sprayed cooling water in the condenser, which is directly connected to the turbine, the wet steam becomes condensed water at 49°C. This condensed water is pumped to the cooling tower by a condensate pump and is cooled to 27°C. Cooled water is used as condenser cooling water, oil cooling water, etc. This cooling water is delivered to the condenser by utilizing potential energy between the cooling tower and the condenser rather than by pumping, and also by vacuum conditions in the condenser interior. Since the condensed water is recycled in this system, no water replenishment is required from the exterior. Overflowing condensed water is fed back to the underground through injection wells. Noncondensed gases contained in the steam are continuously ejected from the condenser by using steam ejectors. Steam flowing through the ejectors amounts to approximately 34 tons per hour, about 4% of the total steam. Since 4,000kW power is consumed for driving the condensate pump and the cooling fans and other pumps, the net power output at high tension side of step-up transformer is 106,000kW.

(ii) ***Geothermal Energy*:** Advantages & disadvantages

Advantages:

Geothermal power plant unit capacity has been increasing in recent years, supported by technological innovation and driven by the economics of larger installations. At the same time, demand for small sized geothermal power plant has also expanded. Small sized geothermal power plant is generally used for the following purposes:

- An experimental unit as a pilot plant for a larger size installation
- Meet the needs of electricity demand in limited area
- Power source during construction
- Auxiliary or emergency power source for main geothermal generating plant
- Simplification steam transmission lines as a well-head unit, because geothermal wells are scattered in geothermal field
    - Renewable as long as water is heated naturally
    - Much lower greenhouse gas emissions than fossil fuels
    - Can be inexpensive in areas where geothermal heating naturally occurs

Disadvantages

- Heated water may give out after a while-hotspot moves or aquifer pressure drops
- Geothermal hot spots are sparsely distributed.
- Geothermal energy is quite low grade because the temperature of steam is only between 150 to 200 c(at 100 psi)
- Salts in water can corrode equipment, shorten lifespan
- Limited to geographic areas where geothermal heating naturally occurs
- Geothermal steam contains up to several percent of gaseous impurities, geothermal steam turbines require much more technical consideration than standard thermal steam turbines.
- accumulation of and erosion by solid substances in the steam paths

1.9 Hydrogen

Plants and animals that lived ages ago have returned to haunt us with a vengeance. Their incinerated remains pollute both land and sea and clog the air we breathe. Life from the past now threatens life of the present. As the environmental destruction associated with man's whole sale consumption of fossil fuels has become globally recognized a corresponding need has grown for an alternate energy source. Easy to produce and non-polluting hydrogen could be the ideal for the future. As a gas, hydrogen could be piped to homes and businesses for heating and cooking purposes or converted into electricity by fuel cells. As a cryogenic liquid, hydrogen electricity by fuel cells. As a cryogenic liquid, hydrogen could launch rocket or fly aircraft or locked as a solid in metal hydride storage canisters, hydrogen solid could propel ground transportation and all this could be provided with virtually no impact on the environment.

Hydrogen is a colorless, odorless gas that accounts for the 75% of the entire Unverse's, mass. Hydrogen is found on Earth only in combination with other elements such as oxygen, carbon and nitrogen. To use hydrogen, it must be separated from these elements.

Hydrogen can be made from molecules called hydro carbons by applying heat, a process known as referring hydrogen. This process makes hydrogen from natural gas. An electric current can also be used to separate water into its compound of oxygen and hydrogen is a process known as electrolysis. Some algae and bacteria using sunlight as their energy source give off hydrogen under contain conditions.

Hydrogen as a fuel is high in energy, yet a machine that burn pure hydrogen produces almost zero pollution. NASA has used liquid hydrogen since 1970's to propel rockets and now the space shuttle into orbit. Hydrogen fuel cells power the shuttle's electric systems, producing a clean by –product pure water, which the crew drinks.

The bad news about hydrogen is that although hydrogen is all around us it is chemically locked up in water and organic compounds such as methane and gasoline. The good news is that we can produce it from something we have in plenty i.e. water. Water can be split by electrolysis (electricity) or high temperature (thermolysis) into hydrogen and oxygen. The major problem is that if takes high to produce hydrogen. There are other ways to produce hydrogen.

One is reforming, in which high temperature and chemical process are used to separate hydrogen from carbon atoms in organic chemicals (hydro carbons) found in conventional carbon-containing fuels such as gas, gasoline or methanol. Gasification of coal or biomass can also produce it. Hydrogen can be stored in compressed gas tanks or in the liquid form (liquid hydrogen).In 2002 scientist were also able to trap hydrogen gas in a frame work of water molecules called castrate hydrates.

(i) ***Hydrogen Energy*****:** Advantages & disadvantages of hydrogen energy is given below

Advantages	**Disadvantages**
• Can be produced from water.	• Not found in nature.
• Low environmental impact.	• Energy is needed to produce fuel
• No carbon dioxide emissions if produced from water.	• Negative net energy.
• Good substance for oil.	• Carbon dioxide emission if produced from carbon containing compounds.
• Competitive price if environmental and social costs are included in cost comparisons	• Non- renewable if generates by fossil fuels or nuclear power.
• Easier to store than electricity.	• High costs.
• Safer than gasoline an natural gasoline.	• Short driving range for current fuel cells cars.
• High efficiency 65-95% in full cells.	• No fuel distribution system in place
	• Excessive hydrogen leaks may deplete ozone.

1.10 Solar Energy

Solar energy is clean and eco-environmental friendly renewable energy resource. It has a low energy density, most suitable from conservation of eco-system point of view. Solar energy (radiant light and heat from the sun) has been harnessed by humans since ancient times using a range of ever-evolving technologies.

The sun has an effective black body temperature (T_s) of 5777 K. It is a sphere of intensely hot gaseous matter with a diameter of 1.39×10^9 m and is, on an average, 1.5×10^{11} m from the earth. The sun is a continuous fusion reactor. The maximum spectral intensity occurs at about 0.48 μm wavelength (λ) in the green portion of the visible spectrum. About 8.73% of the total energy is contained in ultraviolet region ($\lambda < 0.40$ μm); another 38.15% is contained in the visible region (0.40 μm $< \lambda <$ 0.70 μm) and the remaining 53.12% is contained in the infrared region ($\lambda >$ 0.70 μm).Solar radiations while passing through the earth's atmosphere are subjected to the mechanisms of atmospheric absorption and scattering. .The spectrum of solar light at the Earth's surface is mostly spread across the visible and near-infrared ranges with a small part in the near-ultraviolet as shown in Fig. 1.7.

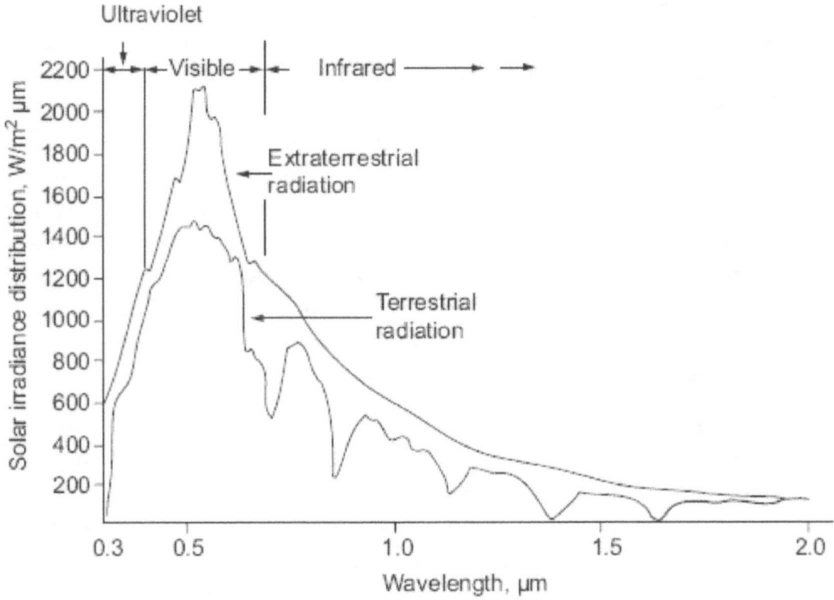

Fig. 1.7 Solar Spectrum

The earth' atmosphere has the unique properties as follows:

It absorbs the ultraviolet and far infrared radiation and allows only radiation having wave length range between 0.29μm to 2.3 μm (short wave length radiation).

It also does not allow radiation having wave length ≥9 μm (long wave length radiation)

The radiant energy flux received per second by a surface of unit area held normal to the direction of sun's rays at the mean earth-sun distance, outside the atmosphere (extra-terrestrial region), is practically constant throughout the year. This is termed as the solar constant I_{sc} and its value is now adopted to be 1367 W/m^2.

The solar radiation, through atmosphere, reaching the earth's surface can be classified into two components: beam radiation and diffuse radiation. Solar radiation propagating along the line joining the receiving surface are referred as *beam/direct radiation* where as solar radiation scattered by aerosols; dust and molecules and does not having a unique direction are called *diffuse radiation.* The total radiation (I) sum of the beam and diffuses radiation and is sometimes referred to as the *global radiation.*

(i) ***Utilization of Solar Energy*:** Solar energy can be utilised directly by two technologies: namely (i) Solar Thermal, and (ii) Solar Photovoltaic. Solar thermal system provides thermal energy for various processes. In cold climate regions, large amount of low grade thermal energy is required to heat air for comfort and hot water for washing, cleaning and other domestic and industrial purposes. Solar collectors are the heart of solar thermal energy systems in which solar energy is collected by absorbing radiation in an absorber and then transferring to a fluid. In general, there are two types of collectors such as flat plat solar collector and concentrating type solar collector.

Flat Plate Collector: The flat-plate collector is one of the most popular and economical solar energy collection system designed for operation in the low temperature range (ambient-60 ^{0}C) or in the medium temperature range (ambient-100 ^{0}C). It absorbs solar energy; convert it into heat and then to transfer absorbed heat to a stream of liquid or gas. It absorbs both the beam and the diffuse radiation, and is usually placed on the top of a building. A flat-plate collector typically consists of (i)Glazing cover (ii) Tubes/evacuated-glazed tube (iii) An absorber plate (iv) Header or manifolds

(v) Bottom glass wool insulation (vi) A container or casing. The whole assembly is fixed on a supporting structure that is installed in a tilted position at a suitable angle facing south in northern hemisphere. For the whole year, the optimum tilt angle of collectors is equal to the latitude of its location. When temperatures higher than 100° C are required, it becomes necessary to concentrate the radiation. Such temperature can be achieved by using focusing or concentrating collectors. It can also be used at high operating temperature only by changing the configuration of absorber by evacuated tube and hence it is referred as evacuated tubular collector. A cross-sectional view of plate flat plate collector is shown in Fig. 1.8.

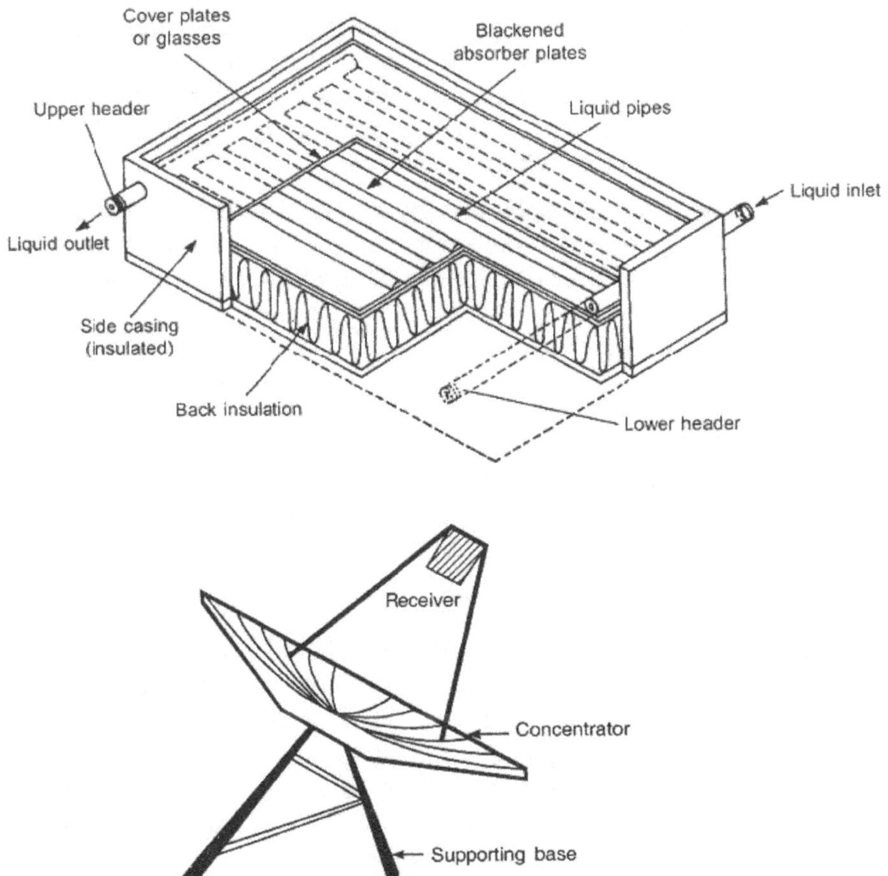

Fig. 1.8 Plate flat plate collector & Parabolic Dish collector

Concentrating Collectors: When temperatures higher than 100° C are required, it becomes necessary to concentrate the radiation. This achieved by using focussing or concentrating collectors. There are some factors that affect the performance of any concentrating type collector like aperture, acceptance angle and concentration ratio. Aperture is opening of the concentrator through which solar radiation passes. Acceptance angle is the angle across which beam radiation may deviate from the normal to the aperture plane and then reach the absorber. Concentration ratio (CR) is the ratio of effective area of the aperture to the surface area of the absorber. The value of CR may change from unity (for flat plate collectors) to thousand (for parabolic dish collector).

(ii) ***Application of Solar Energy*:** Solar Energy can be harnessed using Solar Panels/Collectors. A partial list of solar applications includes space heating and cooling through solar architecture, potable water via distillation, day lighting, solar hot water, solar cooking, high temperature process heat for industrial purposes and electrical power generation through photovoltaics.

In solar thermal route, solar energy is converted into thermal energy with the help of solar collectors and receivers known as solar thermal devices. These devices can be used for different application as:

(i) water heating
(ii) space heating
(iii) power generation
(iv) space cooling and refrigeration
(v) distillation
(vi) drying
(vii) cooking

(a) ***Water Heating*:** Solar water is one of the most common applications of solar energy. A simple layout of solar water heating system with natural circulation is shown in (Fig. 1.9). The system consists of a flat plate collector, normally single glazed, and a storage tank kept at a height. As water in the collector is heated by solar energy, it flow automatically to the top of the water tank and it is replaced by cold water from the bottom of tank (thermosyphon effect). Hot water for use is withdrawn from the top of the tank. An auxiliary electrical emersion heater may be used as back up for use on cloudy or rainy days. When large quantity of hot water is needed, a natural circulation system is not suitable. Large arrays of flat plate collectors are then used and circulation is maintained with water pump (Fig. 1.9).

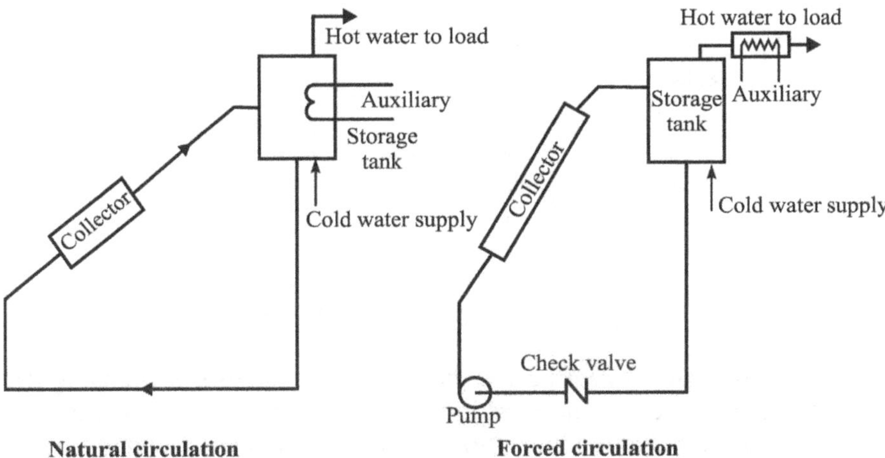

Fig. 1.9 Solar water heating system

(b) ***Space heating*:** Solar energy is also used for passive and active heating a building to maintain comfortable inside. A solar passive space heating system is shown in Fig. 1.10.The south facing wall is made of concrete, adobe, stone or composites of brick block and sand, designed for thermal storage. In order to increase the absorption, the outer surface is painted black. The entire south wall is covered by one or two sheets of glass or plastic sheet with some air gap (10-15 cm) between the wall and inner glazing. Solar radiation after penetration through the glazing is absorbed by the thermal storage wall. The air in the gap between the glazing and the wall thus gets heated, rises up and enters the room through the upper vent while cool air from the room replaces it from the bottom vent. The circulation of air continues till the wall goes on heating the air. Thus, the thermal wall collects stores and transfers the heat to the room.

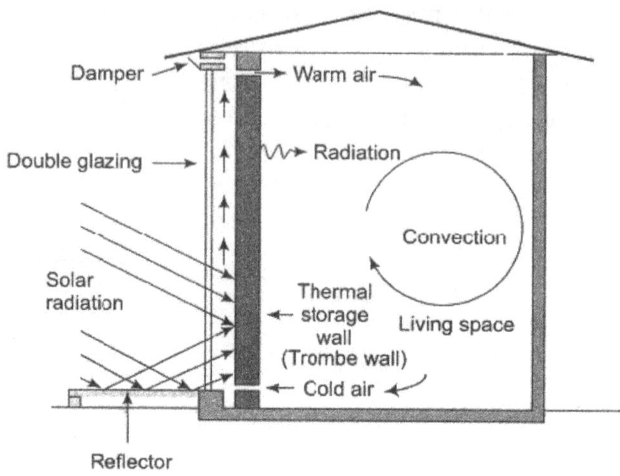

Fig. 1.10 Passive solar space heating

(c) ***Power generation*:** Solar thermal power plants system using flat plate collector and working on Rankine cycle as shown in Fig. 1.11. Hot water (above 90°C) is collected in an insulated tank. It flows through a heat exchanger, through which the working fluid of the energy conversion cycle is also circulated. The working fluid is either methyl chloride or butane having a low boiling temperature up to 90°C. Vapors so formed operate a regular Rankine cycle by flowing through a turbine, a condenser and a liquid pump. The overall efficiency of the generating system is about 2%.

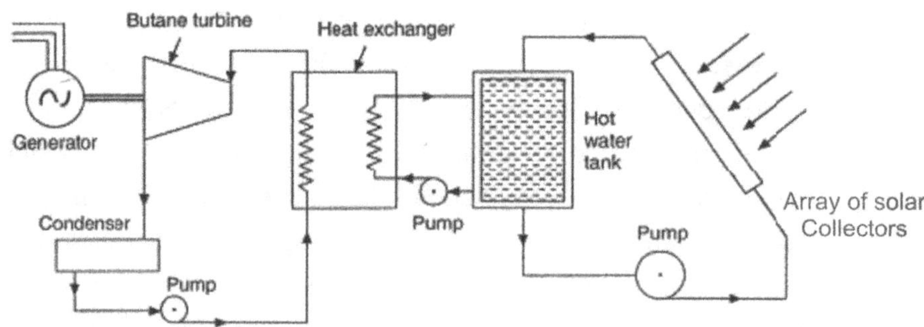

Fig. 1.11 Solar thermal power plant

(d) ***Space cooling and refrigeration*****:** A simple solar operated absorption system is shown in Fig.1.12. Water is heated in a flat plate collector array and passed through a heat exchanger called the generator. Suitable chemical solution for absorption cooling are (NH_3-H_2O where NH_3 is used as the working fluid, and (ii)) LiBr-H_2O solution, where H_2O operates as the working fluid.

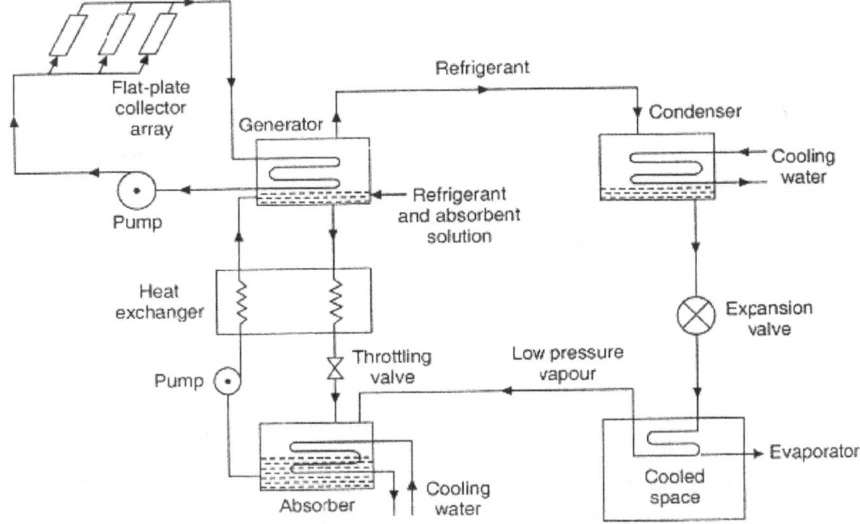

Fig. 1.12 Solar absorption cooling

The whole system consists of four units: generator, condenser, evaporator and absorber. The generator contains a solution mixture of absorbent and refrigerant, and this mixture gets heated with solar energy. Refrigerant vapor is boiled off at a high pressure and flow into condenser, where it gets condensed rejecting heat and becomes liquid at high pressure. Refrigerant then passes through the expansion valve and evaporates in the evaporator. The refrigerant vapor is then absorbed in to a solution mixture taken from generator in which the refrigerant concentration is quite low. The rich solution thus prepared is pumped back to the generator at a high pressure to complete the cycle. A heat exchanger is provided to transfer heat between solution flowing between the absorber and the generator.

(e) ***Distillation*****:** Fig. 1.13 shows the conventional double-slope symmetrical solar distillation unit. It is an airtight basin, usually constructed out of fiber-reinforced plastic with a top cover of transparent material like glass, and plastic. The inner surface of the base known as basin liner is blackened to efficiently absorb the solar radiation incident on it. There is a provision to collect distillate output at lower ends of top cover. The brackish or saline water is fed inside the basin for purification using solar energy. The working principle of the distiller unit is as follows:

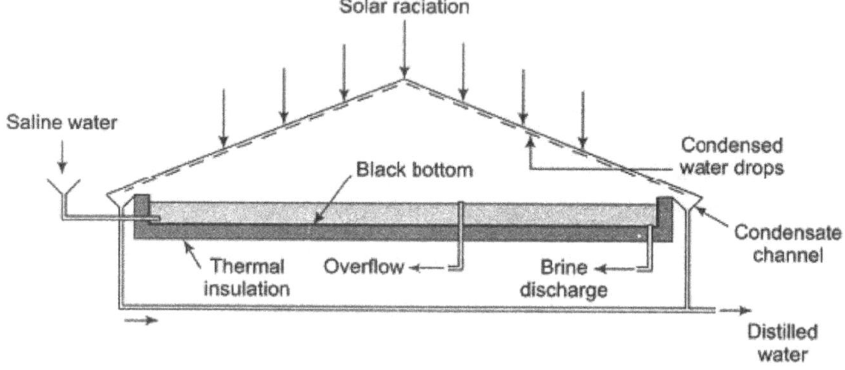

Fig. 1.13 Solar distillation

The solar radiation after reflection and absorption by the glass cover is transmitted inside the enclosure of distiller unit. The transmitted radiation is further partially reflected and absorbed by the water mass. The remaining solar radiation finally reaches the blackened surface where it is mostly absorbed and converted into thermal energy. After absorption of solar radiation at blackened surface (basin liner), most of the thermal energy is convected to water mass and the rest is lost to atmosphere. The water, thus, gets heated, leading to an increase of temperature difference of water and glass cover. The evaporated water gets condensed on the inner surface of the glass cover. The condensed water drop into the channels. The collected distilled water in the channel is taken out of the system for appropriate use.

(f) ***Drying***: There is need of drying of food for storage to avoid food losses between harvesting and consumption. High moisture content is one of the reasons for its spoilage during the course of storage. Drying helps in reducing the moisture content of a product to a level below which deterioration does not take place.

Open sun drying is traditional method of crop drying which has been used since the ancient time. In addition to open sun drying, a more scientific method of solar energy deployment for crop drying has been emerged, which is known as solar drying. Solar cabinet dryer emerged as scientific method of solar energy utilisation for crop drying, which is known as controlled drying or solar drying. Fig. 1.14 shows the principle of direct solar crop drying. This is also referred as cabinet dryer.

In cabinet type solar dryer, a part of incidence solar radiation on the glass cover is reflected back to the atmosphere and remaining enters inside cabinet dryer. Further, a part of transmitted radiation is reflected back in the form of short wavelength from the surface of the crop, which is again transmitted to atmosphere through the glass cover. The remaining part is absorbed by the surface of the crop. Due to the absorption of solar radiation, the crop temperature increases and the crop start emitting long wavelength radiation, which is not allowed to escape to atmosphere due to presence of glass cover unlike open sun drying. Thus the temperature above the crop inside chamber becomes higher. The glass cover serves one more purpose of reducing convective losses to the ambient which further becomes beneficial for rise in crop and chamber temperature. The moisture is taken away by the air entering into the chamber from below and escaping through another opening provided at the top of the dryer due to natural convection.

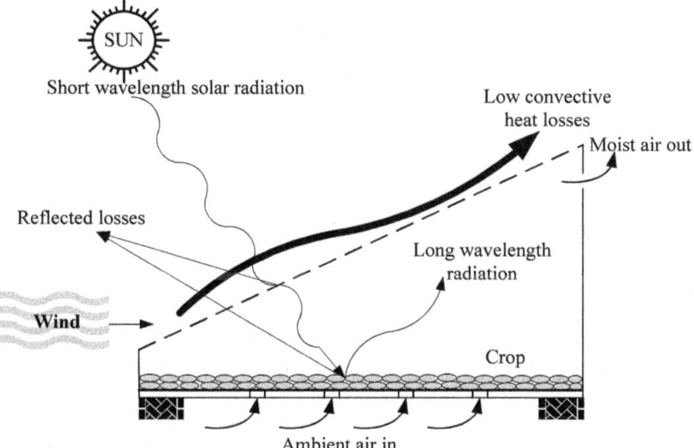

Fig. 1.14 Schematic diagram of direct solar drying (cabinet dryer)

(g) ***Cooking*:** Construction of a most common, box type solar cooker is schematically shown in Fig.1.15. The external dimensions of a typical family size (4 dishes) box type cooker are 60×60×20 cm. This cooker is simple in construction and operation. An insulated box of blackened aluminium contains the utensils with food material. The box receives direct radiation and also reflected radiation from a reflector mirror fixed on inner side of the box cover hinged to one side of the box. A glass cover consisting of two layers of clear window glass sheets serves as the box door. The glass cover traps heat due to the greenhouse effect. Maximum air temperature obtained inside the box is around 140-160°C. This is enough for cooking the boiling type food slowly in about 2-3 hours. It is capable of cooking 2 kg of food.

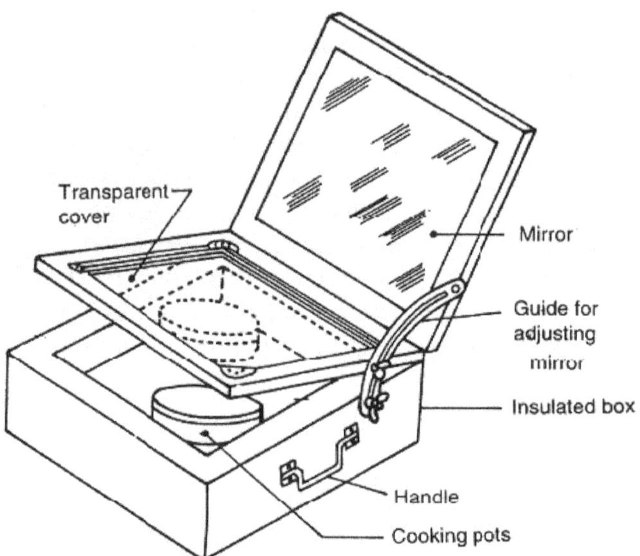

Fig. 1.15 solar cooker

(h) ***Solar Photovoltaic*:**

Solar power can be converted directly into electrical power in photovoltaic (PV) cells, commonly called solar cells.

Sunlight contains photons with energies that reflect the sun's surface temperature; in energy units of electron volts (eV), the solar photons range in energy (hv) from about 3.5 eV (ultraviolet region) to 0.5 eV (infrared region). The energy of the visible region ranges from 3.0 eV (violet) to 1.8 eV (red); the peak power of the sun occurs in the yellow region of the visible region, at about 2.5 eV. At high noon on a cloudless day, the surface of the Earth receives 1,000 watts of solar power per square meter (1 kW/m^2).

(i) ***Working Principle Solar PV Cell*:** Photovoltaic cells generally consist of a light absorber that will only absorb solar photons above certain minimum photon energy. In inorganic semiconductor materials, such as Si, electrons (e-) have energies that fall within certain energy ranges, called bands. The energy ranges, or bands, have energy gaps between them. The band containing electrons with the highest energies is called the valence band. The next band of possible electron energies is called the conduction band; the lowest electron energy in the conduction band is separated from the highest energy in the valence band by the band gap. When all the electrons in the absorber are in their lowest energy state, they fill up the valence band, and the conduction band is empty of electrons. This is the usual situation in the dark.

When photons are absorbed, they transfer their energy to electrons in the filled valence band and promote these electrons to higher energy states in the empty conduction band. There are no energy states between the valence and conduction bands, which is why this separation is called a band gap and why only photons with energies above the band gap can cause the transfer of electrons from the lower-energy-state valence band into the higher-energy-state conduction band. When photons transfer electrons across the band gap, they create negative charges in the conduction band and leave behind positive charges in the valence band; these positive charges are called holes (h^+). Thus, absorbed photons in semiconductors create pairs of negative electrons and positive holes.

In a PV cell, the electrons and holes formed upon absorption of light separate and move to opposite sides of the cell structure, where they are collected and pass through wires connected to the cell to produce a current and a voltage-thus generating electrical power.

The "peak watt" (Wp) rating is the power (in watts) produced by a solar module illuminated under the following standard conditions: 1,000 W/m^2 intensity, 25°C ambient temperature, and a spectrum that relates to sunlight that has passed through the atmosphere when the sun is at a 42° elevation from the horizon.Because of day/night and time-of-day variations in insolation and cloud cover, the average electrical power produced by a solar cell over a year is about 20% of its Wp rating. Solar cells have a lifetime of approximately 30 yr. They incur no fuel expenses, but they do involve a capital cost.

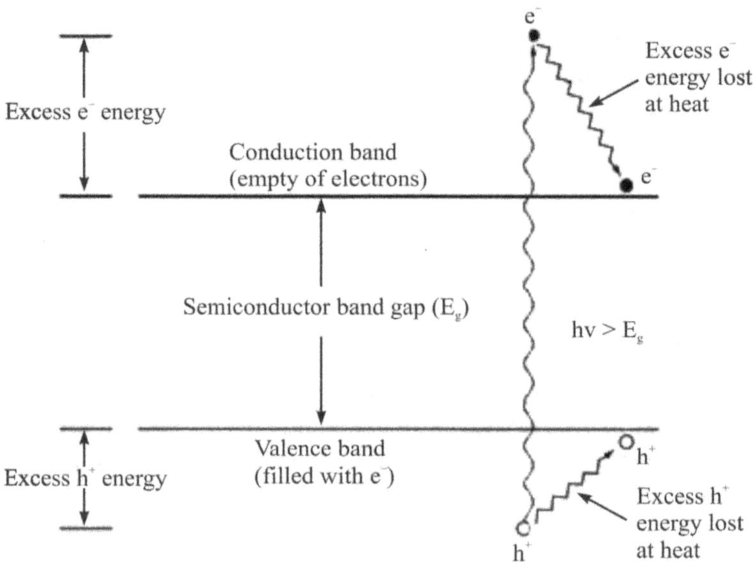

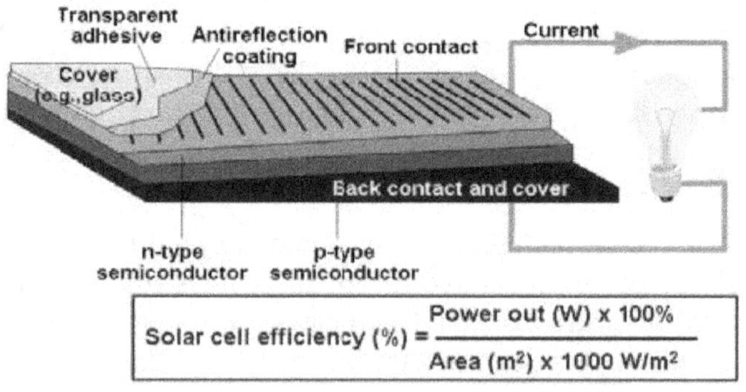

$$\text{Solar cell efficiency (\%)} = \frac{\text{Power out (W)} \times 100\%}{\text{Area (m}^2\text{)} \times 1000\ \text{W/m}^2}$$

10% efficiency = 100 W/m² or 10 W/ft²

Fig. 1.16 Principle of working of solar cell

(ii) ***Types of photovoltaic cells*:** PV cells can be divided into three categories:

1. Inorganic cells, based on solid-state inorganic semiconductors;
2. Organic cells, based on organic semiconductors; and
3. Photo electrochemical (PEC) cells, based on interfaces between semiconductors and molecules.

(iii) ***PV cell, module and array*:** Cells are usually mounted in modules and multiple modules are used in arrays. Individual modules may have cells connected in series and parallel combinations to obtain the desired voltage. Arrays of modules may also be arranged in series and parallel. For identical modules or cells connected in series, the voltages are added, while when connected in parallel, the currents are added (Fig. 1.17).

To protect these cells and modules from damage, a string of cells is hermetically sealed between a layer of toughed glass and layers of ethyl vinyl acetate (EVA). An insulating tidlar sheet is placed beneath the EVA layers to give further protection to the photovoltaic module. An outer frame is attached to give strength to the module and to enable easy mounting on structures. A terminal box is also attached to the back of PV module, where positive and negative ends of solar PV string are welded or soldered to the terminals. This complete assembly constitutes a PV module. PV module can be utilized as a single unit for powering direct or further can be connected to another module which will form an array called PV array.

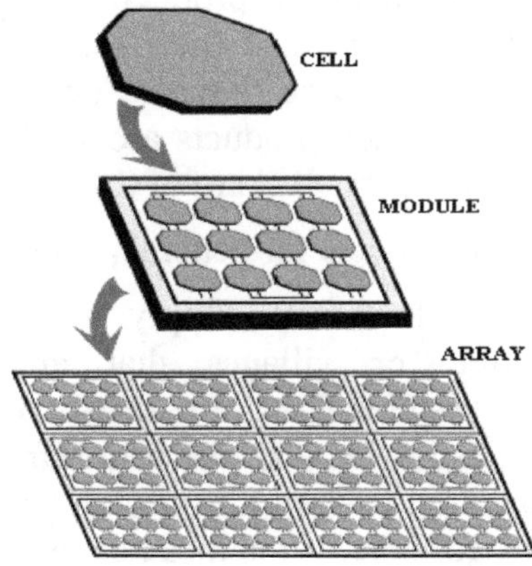

Fig. 1.17 Solar photovoltaic cell, module and array

(iv) ***Solar photovoltaic System*:** Solar photovoltaic module produces DC power. To operate domestic electrical device, needs AC power. Hence, there is a need to convert DC power in to 220 V, 50 Hz AC power. This can be achieving by using invertors. Components other than PV module are collectively known as balance of system (BOS) which includes storage batteries, an electronic charge controller and an inverter. The block diagram of solar photovoltaic system is shown in Fig.1.18.

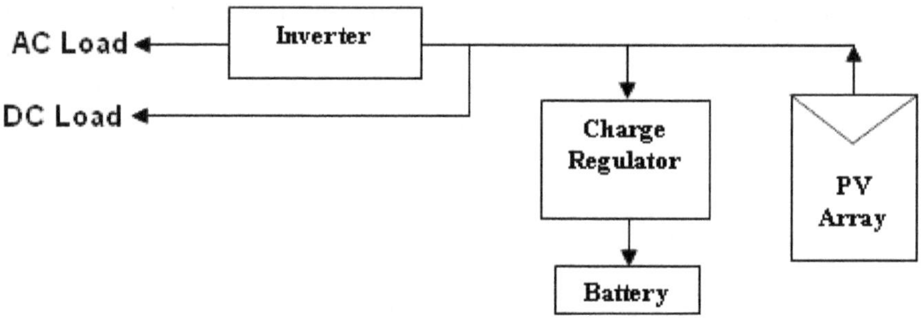

Fig. 1.18 Block diagram of solar photovoltaic system

(v) ***Areas of solar PV application*:** The most important advantages with solar cells are reliability in operation and that the technology can be easily adjusted to a given need in everything from very small systems (fractions of a Watt) to very large plants (MW).

When solar cells are used in small systems (a few Watts or less), they are normally built into a product, for example a street light, fittings for garden lighting and the like. Normally, these systems only provide the one service they are designed for as they don't have a power outlet. Such products are getting more common, but all the same represent a small share of the solar cell market.

There are four system types for general power supply:

Stand-alone systems for private supply, which provide electricity to cabins, households or villages that aren't attached to the transmission grid system. They're normally dimensioned to provide power for lights, radio/TV and possibly refrigerators, and they are used when connection to the transmission grid system is expensive or not technically feasible. If the distance to the system is more than two kilometers, solar cells can be an economic alternative for supplying moderate needs.

Stand-alone systems for other purposes often supply power for specific purposes: telecommunication, water pumping and lighthouses. These systems are used when reliable power supply is needed, it is not possible to establish a grid connection, and it is expensive to provide fuel for generators.

Distributed grid-connected systems are common in a number of countries because of generous subsidy arrangements. Japan and Germany are countries leading the way, and it is these arrangements that have led to the strong growth in the solar cell market. These kinds of systems typically have some kWp solar cells, but can be significantly larger. They reduce the owner's need to buy electricity from the grid. A possible surplus production is sold to the grid.

Centralized grid-connected systems can be of many megawatts and are simply a power plant that uses solar cell technology, since the electricity is fed directly in to a transmission grid.

A solar cell system consists of more components than the solar panels. In stand-alone systems, the most important ones are batteries, charge regulators, cabling and assembling equipment and loads such as lamps and refrigerators.

(iii) ***Solar energy*****:** Advantages & disadvantages of solar are given below

Advantages	Disadvantages
• Solar energy is freely available and in abundance	• Low Flux Density
• Solar energy is pollution free	• Intermittent Availability
• Systems are easy to install, generate and maintain	
• Solar Systems can be specifically designed according to individual requirements.	
• No fuel or moving parts	
• Recurring fuel costs are zero	
• Electricity produced can be stored in batteries and used when it is needed.	

1.11 Wind Energy

Man has harnessed the energy in wind for thousands of years, for sailing boats, powering wind mills and milling grain at land. Among all renewable energy sources, wind power is the most mature in terms of commercial development. The development costs have decreased dramatically in recent years, however most projects are still dependent on public subsidies in order to be profitable. This energy source is interesting because of its renewability and its availability. Wind energy is the kinetic energy associated with movement of large masses of air. These motions results from uneven heating of atmosphere by the sun, creating temperature, density and pressure differences. It is estimated that 1% of all solar radiation falling on surface of the earth is converted into kinetic energy of the atmosphere, 30% of which occurs in the lowest 1,000 m of elevation. It is thus an indirect form of solar energy.

The land has a lower heat capacity than the sea and heats up quickly during day, but at night it cools more quickly than sea. During the day, the sea is therefore cooler than the land and this causes the cooler air to flow shoreward to replace the rising warm air on the land, creating local winds. During the night the direction of air flow is reversed, causing breeze (Fig. 1.19).

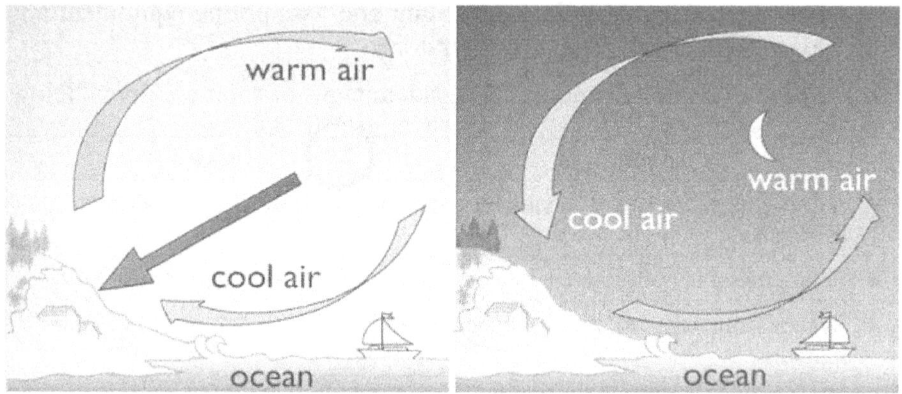

Fig. 1.19 Wind flow in day and night time

On global basis two forces determine the speed and direction of wind. Primary force for global wind is developed due to differential heating of earth at equator and Polar Regions. In the tropical regions there is net gain of heat due to solar radiation, whereas in the Polar Regions there is net loss of heat. This means that the earth's atmosphere has to circulate to transport heat from tropics towards the poles. On a global scale, these atmospheric currents work as an immense energy transfer medium. Ocean currents act similarly and are responsible for about 30% of this global heat transfer.

(i) Energy and Power in the wind

The energy contained in the wind is its kinetic energy. The kinetic energy of any particular mass of moving air is equal to half the mass (m) of the air times the square of its velocity (V):

Kinetic energy = half mass ×velocity squared

i.e. kinetic energy = $0.5\ m\ V^2$ (1)

where m is in kilogram and V is in meters per second (m/s)

We can calculate the kinetic energy in the wind if, first we imagine air passing through a circular ring or hoop enclosing a circular area A (say 100 m^2) at velocity V(say 10 m/s) (Fig. 1.20).

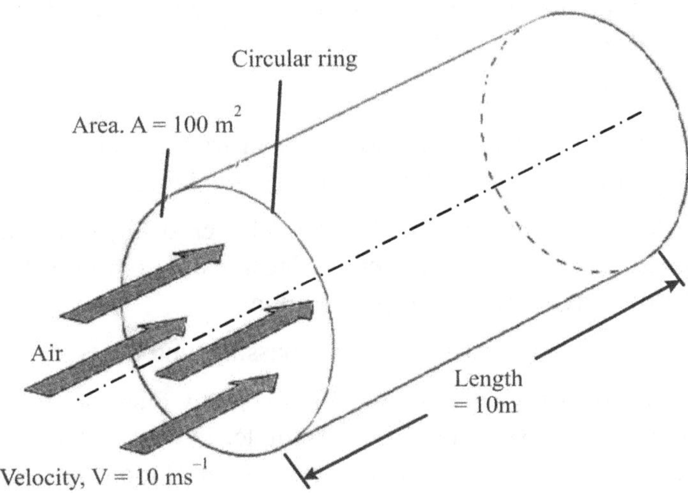

Fig. 1.20 Cylindrical volume of air passing at velocity V through a ring enclosing and area A, each second

As the air is moving at a velocity of 10 m/s, a cylinder of air with a length of 10 m will flow through the ring each second. Therefore, a volume of air equal to 100 ×10=1000 m^3 will pass through the ring each second. By multiplying this volume by the density of air, ρ (which at sea level is 1.2256 kg/m^3), we obtain the mass of the air flowing through the ring per second (in kg/s). In other words:

Mass (m) of air per second = air density × volume of air flowing per second

= air density × area ×length of cylinder of air flowing per second

= air density × area ×velocity

$m = \rho A V$

Substituting for m in Eqn. (1) above:

Kinetic energy per second = $0.5\ \rho\ AV^3$ (joules per second)

where ρ is in kg/m^3, A is in m^2 and V is in m/s

if we recall that energy per unit of time is equal to power, then the power in the wind ,P (watt) = kinetic energy in the wind per second (joules per second)

i.e. $P = 0.5\ \rho\ AV^3$

The main relationships that are apparent from the above calculation are that the power in the wind is proportional to:

The density of the air. The density is lower at higher elevations in mountainous regions; but average densities in cold climate may be up to 10% higher than in tropical regions.

The area through which the wind is passing; and

The cube of the wind velocity. Wind velocity therefore has a strong influence on power output. Fro example, if wind velocity is 4 m/s instead of 3 m/s, power increases by a factor of more that two.

Note that the power contained in the wind is not practice the amount of power that can be extracted by a wind turbine. This is because losses are incurred in the energy conversions process. In addition, some of the air is 'pushed aside' by the rotor and bypass it without generating power. Albert Betz showed in 1928 that the maximum fraction of the power in the wind that can theoretically be extracted is 16/27 (59.3%). This occurs when the undisturbed wind velocity is reduced by one third, in other words, when the axial interference factor is equal to one-third. The value of 59.3% is often referred to as the **Betz limit**. Actual efficiency is generally 50% to 70% of maximum efficiency. Therefore, the actual efficiency is approximately 35%

(ii) ***Classification of Wind Turbine*:**

Lift and Drag type wind turbine: wind turbine can be recognized on their geometry and the approach in which wind passes over the blade. In principle, there are two different types of wind energy conversion devices such as lift and drag.

Low speed devices are mainly driven by the drag forces acting on the rotor. They generally move slower than the wind and their motion reduces rather than enhancing the power extraction. The torque at the rotor shaft is relatively high, which is prime importance for mechanical applications such as water pumps. A greater blade area is essential for slower turbine, so the fabrication of blade is undertaken using curved plates.

High speed turbines rely on lift forces to move the blades of turbine and the linear speed of the blades is usually several times faster than the wind speed. The torque is low as compared to drag type. For electric power generation, the shaft of the generator requires to be driven at a high speed. For the same swept area, the energy extracted by a wind turbine operating on lift forces is several times greater than the energy from drag type turbine. Therefore lift type of devices are preferable to the drag type for electric power generation.

Horizontal axis and vertical axis wind turbine: Modern wind turbines come in two basic configurations: horizontal axis and vertical axis, based on axis over which their rotor revolves. Horizontal axis turbines are predominantly of the axial flow type, whereas vertical axis turbines are generally of the cross flow type.

Horizontal axis turbines are those in which the axis of rotation is horizontal with respect to ground i.e. the rotating shaft is parallel to the ground and the blades are perpendicular to the ground. The vertical axis turbines are those in which the axis of rotation is vertical with respect to the ground. Fig. 1.21 shows the configuration of horizontal and vertical turbines.

These days over 90% of the wind turbines are horizontal axis machine, they are mostly two or three bladed propeller type of machines. Different designs of horizontal axis wind turbines have been shown in Fig. 1.22.

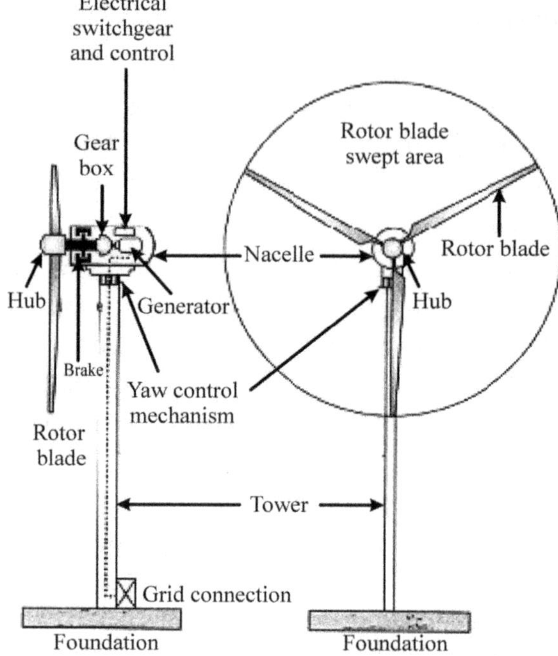

Horizontal axis

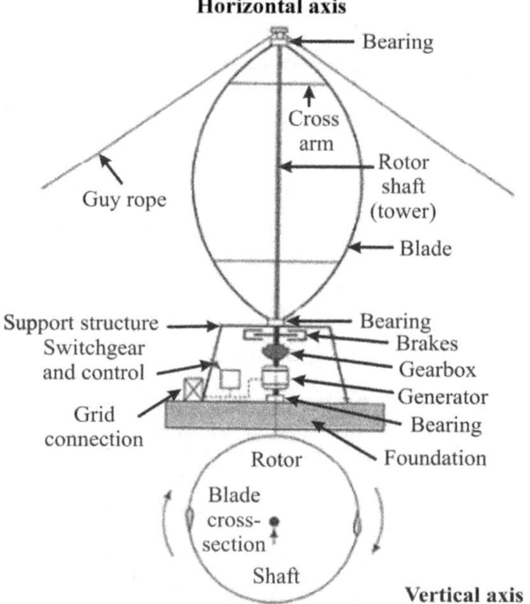

Vertical axis

Fig. 1.21 Wind Turbine Configuration

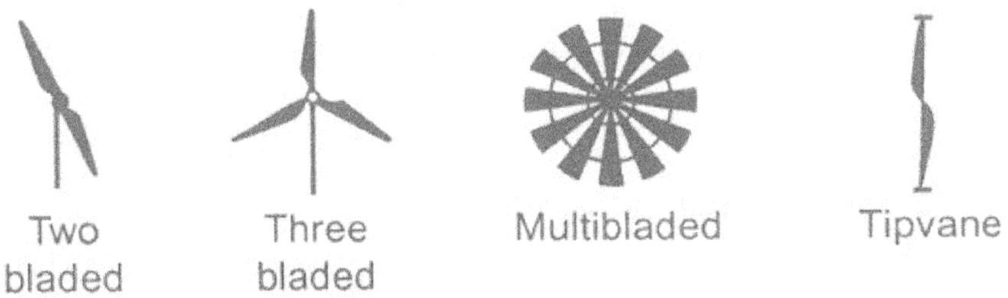

Fig. 1.22 Different design of horizontal axis wind turbine

Vertical axis wind turbine, unlike their horizontal axis counterparts, can harness winds from any direction without the need to reposition the rotor when direction changes. Different types of vertical axis wind turbines have been shown in Fig. 1.23.

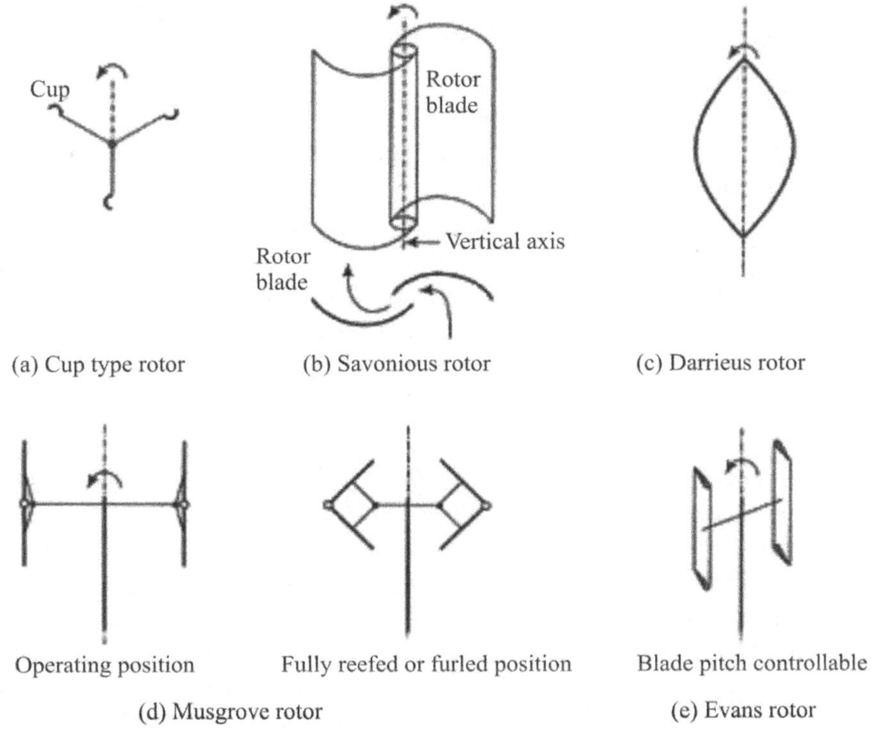

Fig. 1.23 Different design of vertical axis wind turbine

(iii) ***Working of WEC*****:** Components of Wind Energy Conversion System: The general schematic diagram of horizontal axis wind turbine generator is shown in Fig. 1.24. The main components of a wind turbine for electricity generation are as follows:

- Blades
- Nacelle
- Power transmission system
- Generator
- Yaw control
- Brakes
- Controllers
- Tower

Blades: Wind turbine blades need to be lightweight and posses adequate strength and hence require to be fabricated with aircraft industry techniques. The blades are made of glass fiber reinforced polyester with a suitable structural geometrical shapes shape to create lift as air flows over them

Nacelle: The rotor attached to the nacelle, which is placed on the top of the tower. It includes the gear box, low and high speed shafts, generator controller and rotor brake.

Power transmission system: The mechanical power generated by the rotor blades is transmitted to the generator by a two stage gear box located in the nacelle. The gear box is needed to increases the speed of the rotor, from typically 20 to 50 revolution per minute to 100 or 1500 rpm which is required for driving most types of generator.

Generator: A generator converts mechanical motion of the rotor to electrical energy it works on the principle of electromagnetic induction. Generator of grid connected wind machine is required to produce the frequency which is same as grid frequency. Generator can be synchronous or asynchronous generator. Synchronous generator provides constant frequency output power but does not allow rotor speed variation. On the other hand, an induction generator can supply constant frequency output power while allowing some variation in the rotor speed. Practically, wind speed varies and thus variation in rotor speed is expected, induction generators are normally used.

Yaw control: The yaw controls continuously orients the rotor on the direction of the wind. It can be as simple as the tail vane or more complex on modern towers. A horizontal axis wind turbine has a yaw system that turns the nacelle according to the actual wind direction, using a rotary actuator engaging on a gear ring at the top of the tower. A rotor swept area should be moving perpendicular to the direction. The yaw system is also used to cut off the rotor from the wind in case of very strong wind. This is done from pint of view of protecting the machine.

Brakes: Braking of wind energy generator is done by full feathering. An emergency stop activates the hydraulic disc brake fitted to the high-speed shaft of the gear box.

Controllers: Wind energy generators are monitored and controlled by a microprocessor based control unit. A controller monitors the parameters in the nacelle besides controlling the operation of the pitch system. Variations in the blade position are performed by a hydraulic system, which also delivers pressure to the brake system.

Tower: The most common types of tower are the lattice or tubular types constructed from steel or concrete. The modern medium sized and large wind turbines have tubular tower, which allow access from inside the tower to the nacelle during bad weather conditions. The tower is designed to withstand wind

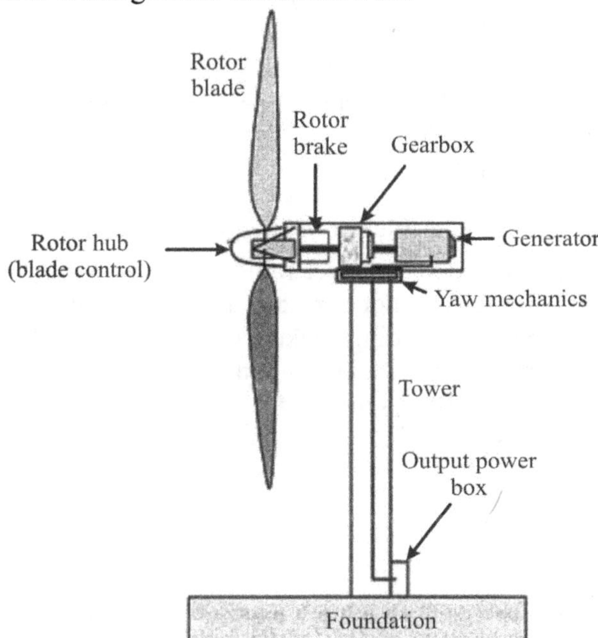

Fig. 1.24 Wind turbine for power generation and its components

(iv) **Criteria for site selection**: Before erecting the wind energy conversion system, it is important to consider the technical, environmental, social, economic factors. Various Studies are conducted on four types of sites such as (i) plane land site (ii) Hill top site (iii) sea shores site and (iv) off-shore shallow water sites, for deciding the locations for wind energy conversion system. The major criteria in selecting the sites for a wind energy conversion system are:

As the building, forest offers the resistance to the air movement, the wind farms are located away from cities and forests.

Flat open area should be selected, as the wind velocities are high in flat open area.

Adequate and uniform average wind velocity throughout the year should be available. The relation between power in the wind and wind velocity is given by

$$P = 0.5\,\rho\,AV^3$$

The power is proportional to the cube of the wind velocity, therefore it is desirable to select a site for wind energy conversion system with high wind velocity.

The readings of wind velocity should be taken at several heights from the ground. For flat open areas the relation between velocity and height is given by

$$V \propto H^{1/7}$$

This relation is applicable for the heights between 50 m to 250 m.

For the site selection it is desirable to have average wind speed. Since wind velocity is not constant, the average velocity should be calculate from the readings taken over a period of one year by average hourly, daily, weekly, monthly readings are suitable. Wind speeds of 4 m/s to 30 m/s are useful for operating range.

$$V_m = \frac{\sum_i^n V_i}{n}$$

where, $\sum_i^n V_i$ = Sum of all velocity observations = $V_1 + V_2 + \ldots\ldots\ldots\ldots V_n$

n = Number of observations

Altitude of the proposed site should be considered. Higher ground experience strong winds than lower ground. Thus, altitude affects the electric power output of wind energy conversion system.

Ground surface should be stable.

If small tree, grass or vegetations are present, then the height of the tower will increases, which increases the cost of the system.

Approach roads upto site for movement of erection equipment structures materials, blades etc.

Site should be near to the consumer, which minimizes the cost and losses

Cost of the site should be favorable.

(iv) ***Wind Power*:** Merits & demerits of wind power is given below

Merits	Demerits
• It is renewable and not depleted with use like fossil fuel.	• Wind energy has low energy density and favorable in geographical locations away from cities and forests.
• It is helpful in supplying electric power to remote areas	• Variable, unsteady, irregular, intermittent, erratic and sometimes dangerous.
• Wind energy generation is eco friendly and no pollution during energy generation.	• Most of the wind energy generation system, using induction generators and due to frequent and sharp change in wind conditions produce various disturbances to the grid system like harmonics, flickering, voltage and current fluctuations.
• Wind power generation id cost effective and reliable.	• Wind energy generation supplies only active power to the grid and consumes reactive power from the grid. Hence, it is necessary to provide appropriate power factor compensation at grid station or wind farm sub-station.
	• Turbine rotors are not very efficient as they extract only 10 to 40% of the available wind energy.
	• Wind energy is capital intensive.
	• Some minor negative impacts are noted like noise, bird hit, land erosion, impact on wild life etc.

(v) Applications of wind energy:

(i) Utility interconnected wind turbines generate power which is synchronous with the grid and are used to reduce utility bills by displacing the utility power used in the household and by selling the excess power back to the electric company.

(ii) Wind turbines for remote homes (off the grid) generate DC current for battery charging.

(iii) Wind turbines for remote water pumping generate 3 phase AC current suitable for driving an electrical submersible pump directly. Wind turbines suitable for residential or village scale wind power range from 500 Watts to 50 kilowatts.

1.12 Biomass Energy

Biomass is a renewable energy resource derived from the solid carbonaceous waste material of various human, animals, plants and natural activities. It is derived from numerous sources, including the by-products from the wood industry, agricultural crops, raw material from the forest, household wastes and discarded material from food processing plants etc. More significantly, it is a store which is continually replenished by the flow of energy from the sun, through the process of photosynthesis.

Biomass does not add carbon dioxide to the atmosphere as it absorbs the same amount of carbon in growing as it releases when consumed as a fuel. Its advantage is that it can be used to generate electricity with the same equipment that is now being used for burning fossil fuels. Biomass is an important source of energy and the most important fuel worldwide after coal, oil and natural gas. Bio-energy, in the form of biogas, which is derived from biomass, is expected to become one of the key energy resources for global sustainable development. Biomass offers higher energy efficiency through form of Biogas than by direct burning .Bioenergy is any energy created from a renewable biological resource.

Biomass is material that comes from plants. Plants use the light energy from the sun to convert water and carbon dioxide to sugars that can be stored, through a process called photosynthesis. Some plants, like sugar cane and sugar beets, store the energy as simple sugars. These are mostly used for food. Other plants store the energy as more complex sugars, called starches. These plants include grains like corn and are also used for food.

Another type of plant matter, called cellulosic biomass, is made up of very complex sugar polymers (complex polysaccharides), and is not generally used as a food source. This type of biomass will be the future feedstock for bioethanol production. Specific feedstocks being tested include agricultural and forestry residues, organic urban wastes, food processing and other industrial wastes, and energy crops. There are many types of biomass resources currently used and potentially available. This includes everything from primary sources of crops and residues harvested/collected directly from the land, to secondary sources such as sawmill residuals, to tertiary sources of post-consumer residuals that often end up in landfills. Biomass resources also include the gases that result from anaerobic digestion of animal manures or organic materials in landfills.

A variety of fuels can be produced from biomass resources including liquid fuels, such as ethanol, methanol, biodiesel, Fischer-Tropsch diesel, and gaseous fuels, such as hydrogen and methane. Biofuels are primarily used to fuel vehicles, but can also fuel engines or fuel cells for electricity generation. Biomass generates electricity by combustion, which releases the stored solar energy contained in the plant matter.3 Unlike wind or solar, a benefit of biomass is that it is "dispatchable" – that is, it can be turned on and off on demand. Utilities in particular like this feature, because it ensures that the power is available when they need it the most.

(i) **Sources of biomass:** Also referred to as "feedstocks", biomass for a power plant can come from a wide variety of sources, including the following:

Wood Residues This refers to leftover wood from other uses, and not wood harvested expressly for biomass. The lumber, pulp, and wood milling industries already extensively use wood waste to produce power. Wood residues can also come from forest thinnings, urban tree trimmings, residual construction material, demolition material, wood pallets, and other waste.

Agricultural Residues This includes primarily mill residues (waste from a processing plant, like nut hulls and oat hulls) and field residues (left in field after harvest, like corn stover and wheat straw). The removal of field residues for energy must be balanced with the benefit to soil quality that residues provide.

Energy Crops These are crops that are "dedicated" for energy production. The most promising include woody crops like willows, hybrid poplars, maple, and sycamore; and herbaceous crops like switchgrass and other prairie grasses. Often these types of energy crops offer environmental benefits over conventional crops, like less need for crop inputs, habitat for wildlife and reduced erosion and run-off.

Animal Waste Dry animal waste, primarily from poultry, can be burned directly for heat and power. Wet manure can be digested to produce biogas.

Sewage Sludge Although the solids can be burned, a more common option for producing energy at a sewage treatment plant is anaerobic digestion, which produces energy while treating the waste

Biofuels Liquid fuels like ethanol and biodiesel are primarily used in transportation applications, but could also be burned to produce electricity.

(ii) ***Biomass to energy***: Biomass gasification is considered one of the most promising routes for syngas or combined heat and power production because of the potential for higher efficiency cycles. Fig. 1.25 shows a schematic presentation of processes involved in biomass gasification.

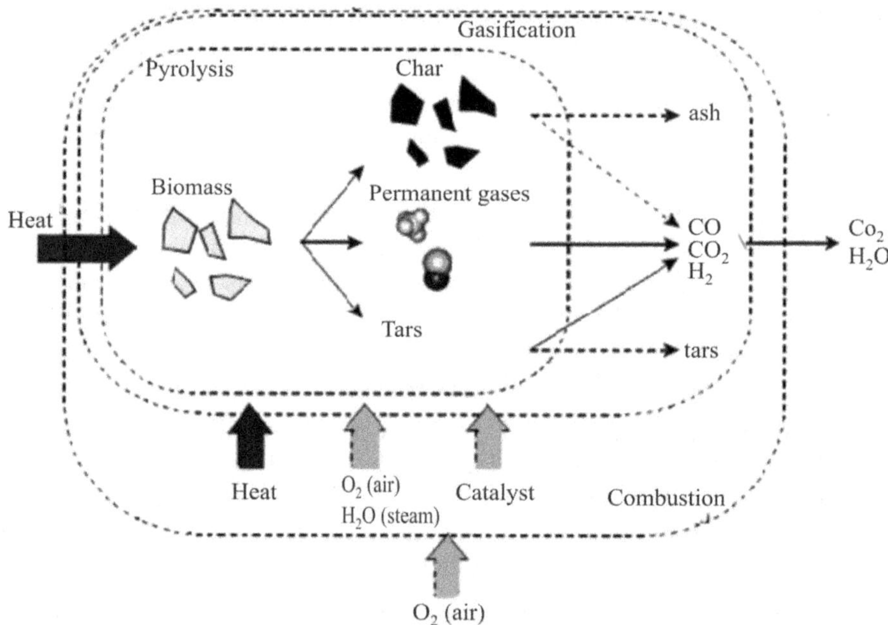

Fig. 1.25 Schematic presentation of gasification as one of the thermal conversion processes

Biomass energy is an important source of energy for majority of the world's population. The use of biomass energy is expected to increase in the near future, with growth in population. In many developing countries, traditional biomass energy dominates national energy statistics, leading to significant negative impacts on human health and the environment. There are, however, opportunities for developing improved and modern biomass energy technologies, which offer substantial benefits in terms of enhanced quality of energy services and reduction in negative health and environmental impacts. In addition, the sustainable harvesting of biomass resources is essential for ensuring the continued availability of this important energy source particularly for the world's poor.

(iii) Biomass Energy: Advantages & disadvantages of biomass energy are given below

Advantages	Disadvantages
• Renewable or recyclable energy source (Stored solar energy)	• Low conversion efficiency of light in to bio mass by plants
• Less waste directed to landfills.	• Relatively low concentration of biomass per unit area of land and water
• Decrease reliance on imported energy sources.	• Scarcity of additional land for growing plant
• Potential rural development and job creation.	• High moisture content of biomass (50%-90%)
• Can generate renewable electricity when the Sun is not shining and the wind is not blowing.	

1.13 Hydroelectrical Energy

Electricity produces from waterpower is known as hydroelectric energy. The potential energy of falling water captured and converted to mechanical energy by water wheel powered the start of industrial revolution. Wherever head or change in elevation could be found, river and stream were dammed and mills were built.

(i) Types of Hydro-Power

Large Scale Hydro power: in this case a high dam is built across a large river to create a reservoir, water is allowed to flow to through huge pipes laid along the steep hill slopes (falling) at controlled rates, thus spinning turbines (prime movers) and in turn generators producing electricity.

***Small hydropower*:** In this case a low dam with no reservoir (or only a small one) is built across a small stream and the water used to spin turbine to produce electricity.

***Pumped Storage hydropower*:** In this case the surplus electricity conventional power plant is used to lift water from a lake or tail race to another reservoir at a higher elevator, water in the upper reservoir is released to spin the turbine for generating electricity.

In 2001, hydro power supplied about 7% of the world's total commercial energy, 20% of the world's electricity. It supplies 99% of the electricity in Norway, 75% in New Zealand and 50% in developing countries and 25% in China. In India the generation of hydro electricity has been emphasized right from the beginning of the First Five Year plan. By the end of Fourth plan, India was able to generate 6.9 thousand MW of hydro electricity, contributing 42% of the total power generation capacity. But due to increase in demand, by the end of Eighth plan it fell down to 25% only. The hydropower potential of India is estimated to be 4×10^{11} k w –hours. Till now we have utilized only a little move than 11% of this potential. Because of increasing concern about the harmful environment and social consequences of large dams, there has been growing pressure on the World Bank and other development agencies to stop funding new large scale hydro power projects.

According to a study by world commission on Dams, hydropower in tropical countries is a major emitter of green house gases. This occurs because reservoirs that power the dams can trap rotting vegetation, which can emit green house gases such as Carbon dioxide and Methane.

Small-scale hydropower projects eliminate most of the harmful environmental effects of large-scale projects. However their power output can vary with seasonal changes in the stream flow.

(ii) *Hydropower*: Following are the advantages of and disadvantages of using large-scale hydropower plants to generate electricity

Advantages	Disadvantages
• Moderate to high net energy	• High construction cost
• High efficiency (80%)	• High environmental impact
• Low cost electricity	• High carbon dioxide emission from biomass in shallow tropical reservoirs.
• Long life span	• Floods natural areas
• No carbon dioxide emission during operation	• Coverts land habitat to Lake Habitat.
• May provide flood control below dam	• Danger of collapse
• Provides water for year-round Irrigation.	• Uproots People
• Reservoir is useful for fishing and recreation.	• Decreases fish harvest Below dam
	• Decreases flow of natural fertilizer (stilt) to land below dam.

According to the United Nations, only about 13% of the World's exploitable potential for hydropower has been developed. Much its untrapped potential is in South Asia, (China), South America and parts of Russia.

1.14 Nuclear Energy

Nuclear energy is non- renewable source of energy, which is released during fission (disintegration) or fusion (union) of selected radioactive materials. Nuclear power appears to be the only hope for large scale energy requirements when fossil fuels are exhausted. The reserves of nuclear fuels is about ten times more than fossil fuels and its major advantage is that even small quantities can produce enormous amounts of energy

Nuclear energy is a nearly carbon-free source of heat and electricity. When it replaces coal-fired power, the operation of a 1-GW nuclear power plant can avoid about 6-7 million tonnes of CO_2 per year, as well as related airborne pollutants. A new generation of safer, more cost effective reactors is being developed. Nuclear energy is harnessed from one of two types of nuclear reactions, fission and fusion.

(i) ***Nuclear Fission*:** Nuclear fission is the process of splitting the nucleus of a heavy atom (target nucleus) into two or more lighter atoms (fission products) when the heavy atom absorbs or is bombarded by a neutron. A few radio nuclides can also spontaneously fission. Fission releases a large amount of energy along with two or more neutrons. For example, a ton of uranium – 235 can produce an energy equivalent 3 million tones of coal or 12 million barrels of oil The large amount of energy released is due to sum of the masses of the fission products being less than the original mass of the heavy atom.

Nuclear energy has been successfully used in the generation of electricity in spaceships, marine vessels, chemical and food-processing industry. The energy from these nuclear reactions is used to heat water in the reactor and generates steam to drive a stream turbine. High temperature gas-cooled reactors and Fast Breeder reactors convert non fissionable Pu ^{239} and U^{233}.

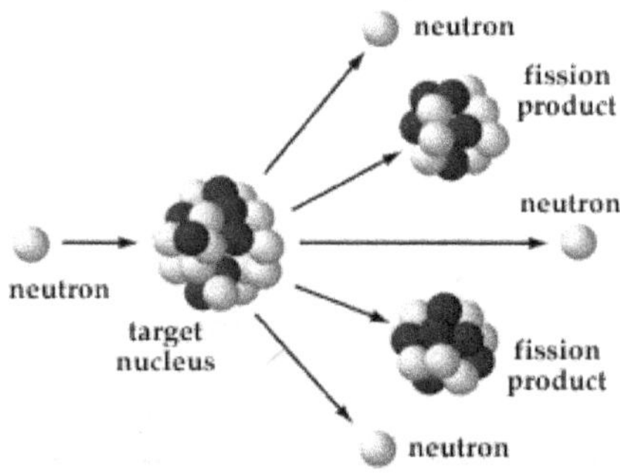

Fig. 1.26 Nuclear fission

In the process, the uranium nucleus splits up into two medium sized nuclei upon neutron collision. Later it was found that there are a few uranium and plutonium isotopes which can undergo fission if they interact with neutrons. These isotopes reach an energetically more favourable state after fission, i.e. more energy is released than the amount that was necessary to cause the fission. Naturally occurring uranium contains three isotopes, U^{234} (0.006%), U^{235} (0.711%) and U^{238} (99.283%). Among these isotopes only U^{235} is fissile i.e very unstable, which undergoes spontaneous fission when bombarded by slow neutrons. The other isotopes U^{238} is a fertile material which requires fast (high energy) neutrons to split the stable nucleus. The number of neutrons released in U^{238} is too small to trigger a chain reaction. Hence for nuclear fission, U^{238} is fertile material that needs conversion into a fissile material.The fission reaction of U^{235} when bombarded by slow neutrons releases fission products neutrons and a large quantity of heat energy (8.2×10^7 kJ per gram of U^{235}, Fig. 1.26. This energy released during fission is equal to 22.78 MWh, while the amount of energy released on combustion of 1 g of coal is only 7 Wh.

(ii) ***Nuclear fusion*****:** It is based on deurterium-deuterium and deuterium-tritum reaction nuclear fusion is also known as thermo nuclear reaction.

Fusion occurs when light atomic nuclei are forced close enough together that they combine to form heavier nuclei, releasing energy in the process. This process powers the sun and stars. Fusion is essentially the reverse of fission, where heavy nuclei break into lighter fragments - the principle behind today's nuclear power plants. To utilise fusion reactions as energy source it is necessary to heat a large volume of dilute gas, containing equal parts of deuterium and tritium to temperatures in excess of 100 million degrees-several times hotter than the centre of the Sun. At these temperatures, the gas becomes plasma. The deuterium and tritium ions in the plasma fuse together to form helium and high speed neutrons, releasing significant amounts of energy. The easiest fusion reaction to recreate in the laboratory is that of deuterium (D) and tritium (T), which are heavy forms of hydrogen. This produces an alpha particle and a neutron, as shown below.

$D+T \rightarrow {}^4He + n + 17.58$ MeV,

$D + D \rightarrow {}^3He + n + 3.27$ MeV,

$D + D \rightarrow T + p + 4.03$ MeV,

$D + {}^3He \rightarrow {}^4He + p + 18.35$ MeV

$p + {}^{11}B \rightarrow 3\,{}^4He + 8.7$ MeV

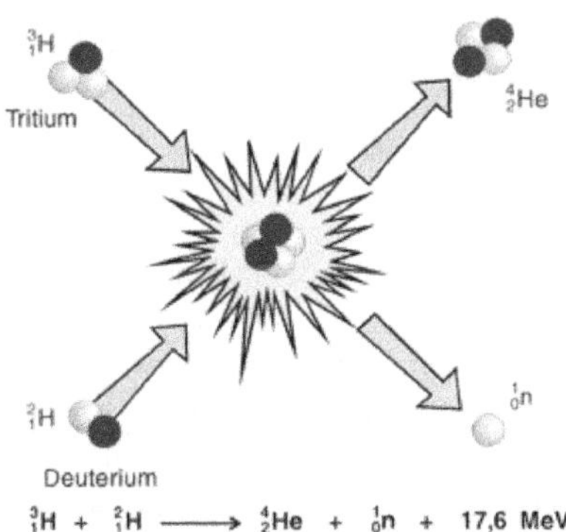

Fig. 1.27 Generation of energy during fusion reaction

The fusion of one kilogram of D-T fuel releases thousands of times more energy than burning one kilogram of coal and has no associated greenhouse gas emissions.

Advantages of fusion:

- Abundant Supply of Fuel (deuterium and tritium)
- No Risk of Nuclear Accident
- No reactor meltdown possible
- Large uncontrolled release of energy impossible
- Minimal or No High Level Nuclear Waste
- Careful material selection should minimize waste caused by neutron activation
- No Air Pollution of Greenhouse Gases Reaction product is Helium and neutron

(iii) *Nuclear Power Plant*:

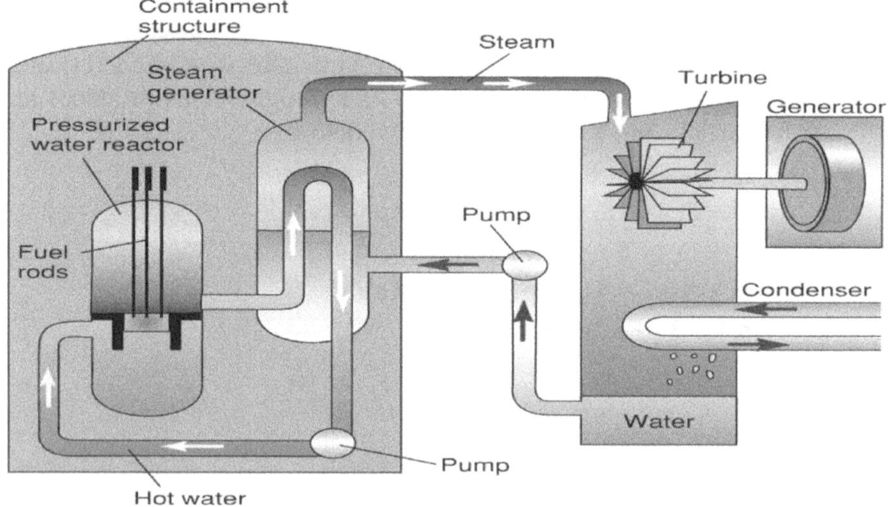

Fig. 1.28 Nucelar Power Plant

Inside a nuclear reactor, the energy released by uranium (U) fission reactions provides heat to a coolant fluid that drives a turbine-powered electricity generator. The fluid may directly drive the turbine or provide heat to a secondary coolant. Nuclear reactors are classified by their neutron energy level (thermal or fast reactors), by their coolant (water, gas, liquid metall), also by their neutron moderator (water, heavy water 2, graphite). Most existing plants are thermal reactors using pressurised (PWR) or boiling water (BWR) as a coolant moderator. Some reactors use heavy-water (PHWR). In the past years, fast-breeder reactors (FBR) have received renewed attention because their fast neutrons can convert U^{238} into Pu^{239} (a usable fuel) and produce fuel in excess of their own consumption. This could increase by 30 times or more the energy extracted from natural U. There is also interest in high-temperature gas-cooled reactors (HTGR). They offer high efficiency, safety, small size and modularity to adapt to different needs and electricity grids.

Beside U^{238}, there is another naturally occurring fertile material thorium (Th^{232}). Both these fertile materials, U^{238} and Th^{232}, can be converted to fissile material plutonium 239 (Pu^{239}) and uranium 233 (U^{233}) respectively, by neutron bombardment.

The neutrons so released are able to fission more uranium atoms. The speed of neutrons is reduced to a critical value by a moderator (graphite or heavy water) for sustained chain reaction.

(iv) ***Nuclear Energy production in India*:** Nuclear power is the fourth-largest source of electricity in India after thermal, hydro and renewable sources of electricity. As of 2008, India has 17 nuclear power plants in operation generating 4,120 MW while 6 other are under construction and are expected to generate an additional 3,160 MW. Since early 1990s, Russia has been a major source of nuclear fuel to India. Due to dwindling domestic uranium reserves, electricity generation from nuclear power in India declined by 12.83% from 2006 to 2008. India now envisages increasing the contribution of nuclear power to overall electricity generation capacity from 4.2% to 9% within 25 years. In 2010, India's installed nuclear power generation capacity will increase to 6,000 MW. As of 2009, India stands 9th in the world in terms of number of operational nuclear power reactors.

1.15 Ocean Energy Sources

Ocean tides and waves contain enormous amounts of energy that can be harnessed. Three sources from oceans:

Tidal power: The twice-daily flow of tides (rising and falling of seas due to the moon's gravitational pull) creates energy of motion that can be converted to electricity.

***Wave power*:** Motion of waves at ocean shores creates energy of motion that can be converted to electricity.

***Ocean Thermal Energy Conversion (OTEC)*:** Exploits differences in warm and cold water. Not yet commercially developed.

Tidal Energy consists of generating kinetic energy from potential energy. If falling water is forced through ducts with rotors attached to them, the rotors will turn driving electric generators .Tide flows through turbines, creating electricity. Generating electricity from tides is very similar to hydroelectric generation, except the tides flow in two directions rather than one. It requires a high tide/low-tide differential of several meters. For tidal power, the most common generating system is the ebb generating system. In the scheme, a dam, or barrage is constructed across an estuary. The tidal basin is allowed to fill when the sluice gates are opened and high tide is in. The gates are then closed when the tide turns trapping the water behind the gates. Once low tide is reached, the gates are opened the water flows through the turbines located underneath the water generating electricity.

In order to create enough electricity to be economically feasible, the size and configuration of the structure has to be increased tremendously. The schematic below shows the basic concept used in an ebb generation scheme (Fig 1.29).

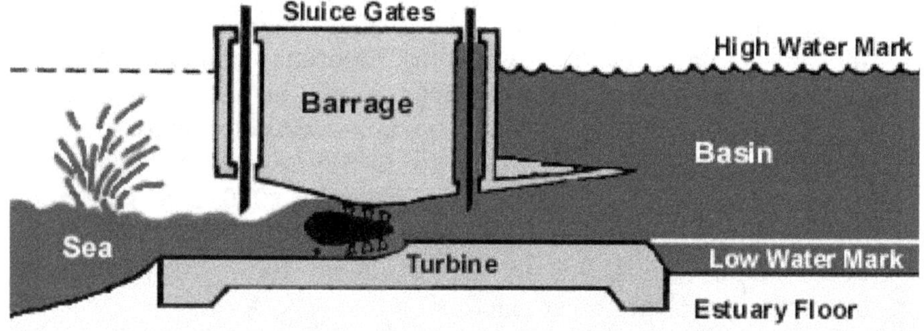

Fig. 1.29 Tidal Power plant

In some cases, double effect turbines are used, which are able to generate electricity when then basin is filling. In this scheme, sluice gates located on either side of the turbine are opened, when the tidal basin is low, and the sea is at high tide level. Water will rush into the tidal basin, turning the turbines and generating electricity. This occurs until the water level on either side of the barrage is equal. At this point, the sluice gates are closed until the sea is at its low tide height. When this occurs, the gates are opened and water flows from the basin to the sea, generating electricity a second time.

Advantages	Disadvantages
Renewable as long as oceans behave as they always have	Development could take up large portions of coastline valuable for other uses
Does not use up large land masses like solar or wind	Could interfere with ecology of estuaries and inter tidal shorelines
No greenhouse gas emissions	OTEC not yet commercially feasible
Much greater power is concentrated in waves than in air.	
Renewable as long as oceans behave as they always have.	

Review Questions

1. Define Environmental studies.
2. Mention the scope and importance of Environmental studies.
3. What are the merits of using renewable energy resources?
4. Why alternate energy resources are required?
5. What is the application of wind energy?
6. Write a note on Geo-Thermal energy.
7. Differentiate between renewable & non-renewable energy resources.
8. Mention the advantages & disadvantages of modern agriculture.
9. What is nuclear energy?
10. What are conventional energy resources? Discuss about solar energy and Ocean thermal energy.
11. Write short notes of (i) Tidal energy (ii) Bio-gas (iii) Nuclear energy.
12. Explain briefly the various methods of harvesting solar energy. Explain the consequences of over utilization of surface and ground water.
13. Discuss the different types of energy sources.

14. Explain the non-renewable sources available to meet the growing energy needs of India. (Solar, Wind, Biomass, Geothermal and Tidal Energy)
15. Explain how the alternate energy sources play an important role in Environmental impact.
16. Write brief on Hydrologic cycle.

CHAPTER 2

Ecosystem

Ecosystem – Segments of Environment: Atmosphere, Hydrosphere, Lithosphere, Biosphere. Cycles in Ecosystem – Water, Carbon, Nitrogen. Biodiversity: Threats and Conservation, Food Chain

2.1 Segments of Environment

Environment can be defined as something that surrounds us. Though our primary interest is the environment of man, we cannot exist in isolation. Human activity has to be understood in relation to other forms of life that exists in both animal and plant kingdom. Therefore it is necessary to deal with the environment of all life forms.

Environment consists of three domains. viz, gaseous – air (atmosphere), liquid – water (Hydrosphere), solid – land (lithosphere) and biosphere.

2.2 Atmosphere

The following points highlight the vital role played by atmosphere in the survival of life in this planet.

- The atmosphere is the protective blanket of gases which is surrounding the earth. It protects the earth from the hostile environment of outer space.
- It absorbs IR radiations emitted by the sun and reemitted from the earth and thus controls the temperature of the earth.
- It allows transmission of significant amounts of radiation only in the regions of 300 – 2500 nm (near UV, Visible, and near IR) and 0.01 – 40 meters (radio waves). i.e it filters tissue damaging UV radiation below 300 nm.
- It acts as a source for CO_2 for plant photosynthesis and O_2 for respiration
- It acts as a source for nitrogen for nitrogen fixing bacteria and ammonia producing plants.
- The atmosphere transports water from ocean to land

(i) ***Composition of atmosphere*:** This gaseous composition of the atmosphere is usually expressed by percentage volume, that is, each gas's relative part of the total mixture. For example, 78% of the atmosphere is made of the gas nitrogen (N_2), 21% is composed of oxygen (O_2), and .9% is made up of argon (Ar). These three gases together make up 99.9% of the atmosphere. Other "vapors" or gases that make up the atmosphere include water vapor (H_20), Carbon Dioxide (CO_2), Neon (Ne), Helium (He), Methane (CH_4), Krypton (Kr), and Hydrogen (H_2). These gases, along with many others, are referred to as "trace" gases, in that there are small traces of them in the atmosphere. The concentration of gases in the atmosphere is measured in parts per thousand (ppt), parts per million (ppm) and parts per billion (ppb).

(ii) ***Layers of atmosphere*:** Most of the gaseous constituents are well mixed throughout the atmosphere. However, the atmosphere itself is not physically uniform but has significant variations in temperature and pressure with altitude. Fig. 2.1 shows the structure of the atmosphere, in which a series of layers is defined by reversals of temperature. The lowest layer, often referred to as the lower atmosphere, is called the troposphere. It ranges in thickness from 8km at the poles to 16km over the equator, mainly as the result of the different energy budgets at these locations. Although variations do occur, the average decline in temperature with altitude (known as the lapse rate) is approximately 6.5 ^{o}C per kilometre. The troposphere contains up to 75% of the gaseous mass of the atmosphere, as well as nearly all of the water vapour and aerosols, whilst 99% of the mass of the atmosphere lies within the lowest 30 km. Owing to the temperature structure of the troposphere, it is in this region of the atmosphere where most of the world's weather systems develop. These are partly driven by convective processes that are established as warm surface air (heated by the Earth's surface) expands and rises before it is cooled at higher levels in the troposphere.

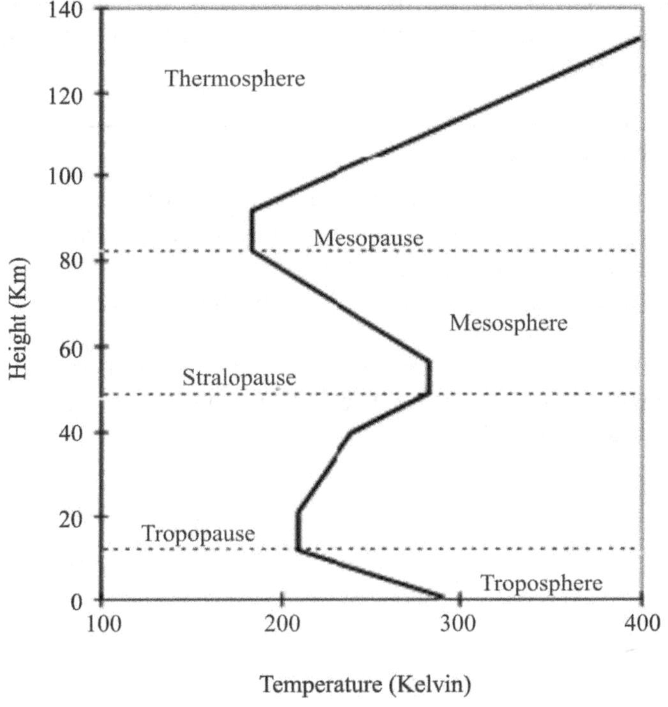

Fig. 2.1 Structure of the Atmosphere

The tropopause marks the upper limit of the troposphere, above which temperatures remain constant before starting to rise again above about 20 km. This temperature inversion prevents further convection of air, thus confining most of the world's weather to the troposphere.The layer above the tropopause in which temperatures start to rise is known as the stratosphere. Throughout this layer, temperatures continue to rise to about an altitude of 50 km, where the rarefied air may attain temperatures close to 0 °C. This rise in temperature is caused by the absorption of solar ultraviolet radiation by the ozone layer Such a temperature profile creates very stable conditions, and the stratosphere lacks the turbulence that is so prevalent in the troposphere.

The stratosphere is capped by the stratopause, another temperature inversion occurring at about 50 km. Above this lies the mesosphere up to about 80 km through which temperatures fall again to almost 100 °C. Above 80 km temperatures rise continually (the thermosphere) to well beyond 1000 °C, although owing to the highly rarefied nature of the atmosphere at these heights, such values are not comparable to those of the troposphere or stratosphere.

2.3 Hydrosphere

The hydrosphere is a collective term given to all different forms of water. It includes all types of water resources such as oceans, seas, rivers, lakes, streams, reservoirs, glaciers and ground waters.97% are ocean water with high salt content and unusable for human consumption.2% of water is locked in polar icecaps and only 1% of the total water supply is available as fresh water in the form of rivers, lakes, streams and ground water for human consumption and other uses. The distribution of earth's water supply is shown in Fig. 2.2.

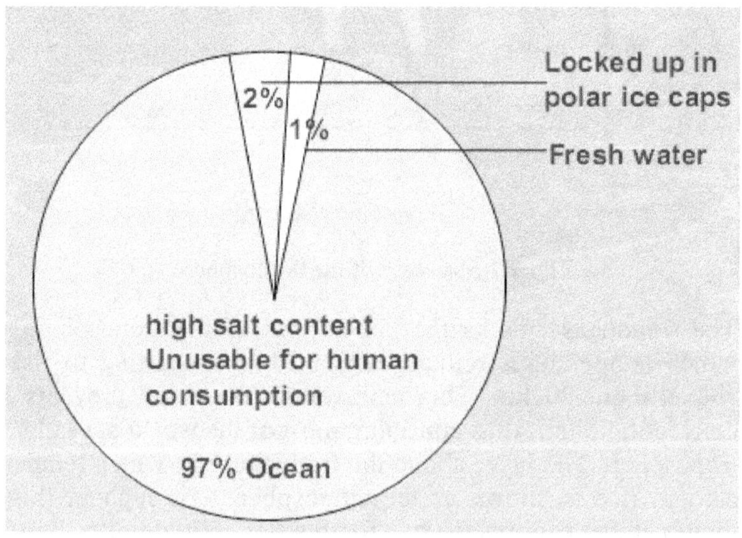

Fig. 2.2 Distribution of earth's water supply

2.4 Lithosphere

- The earth is divided in to layers as shown in Fig. 2.3.
- The lithosphere consists of upper mantle and the crust.

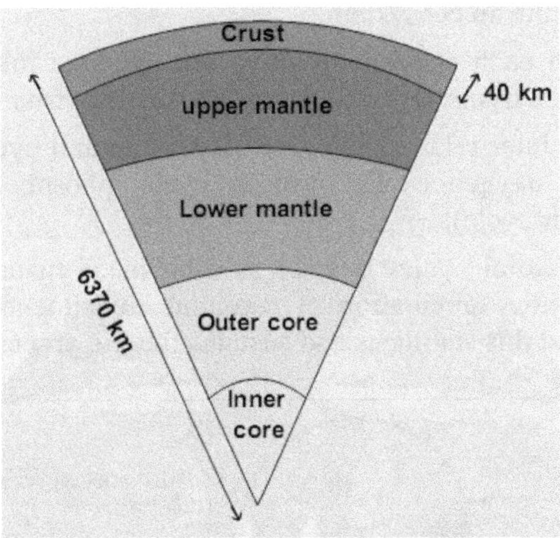

Fig. 2.3 Layers of earth

The crust is the earth's outer skin that is accessible to human. The crust consists of rocks and soil of which the latter is the important part of lithosphere.

2.5 Biosphere

These three domains meet at a common interface on the surface of the earth. This interface, a shallow life-bearing layer is the 'Bio-Sphere'. Structure and functioning of the bio sphere is essentially dependent on the exchange of matter and energy that takes place continuously amongst the land surfaces, water bodies and atmosphere. The biosphere is the global sum of all ecosystems. It can also be called the zone of life on Earth. From the broadest biophysiological point of view, the biosphere is the global ecological system integrating all living beings and their relationships, including their interaction with the elements of the lithosphere, hydrosphere and atmosphere.

- The biosphere refers to the realm of living organisms and their interactions with the environment (VIZ: atmosphere, hydrosphere and lithosphere).

- The biosphere is very large and complex and is divided into smaller units called ecosystems.
- Plants, animals and microorganisms which live in a definite zone along with physical factors such as soil, water and air constitute an ecosystem.
- Within each ecosystems there are dynamic inter relationships between living forms and their physical environment
- These inter relationships manifest as natural cycles.(hydrologic cycle, oxygen cycle, nitrogen cycle, phosphorous cycle and sulphur cycle),
- The natural cycles operate in a balanced manner providing a continuous circulation of essential constituents necessary for life and this stabilizes and sustains the life processes on earth.

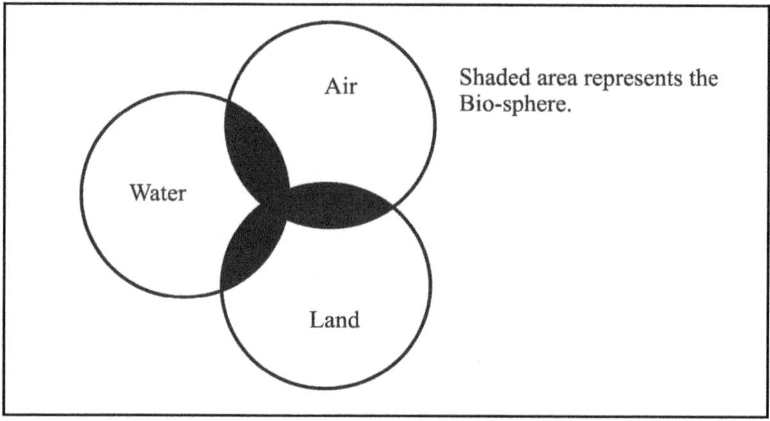

Fig. 2.4 Bio-sphere

2.6 Ecosystem

The bio-sphere is made up of the living system consisting of plant and animal kingdom, and the non-living components including minerals, water etc. The entire system is sustained by the source of energy, the Sun. Organisms belonging to different species either of plant kingdom or animal kingdom interact among themselves as well with the physical environments they occupy. This system is called ecological system or ecosystem.

2.6.1 Balanced Ecosystem

An ecosystem is made up of different components. In the natural environment a balance or equilibrium exists among various organisms and abiotic components. This condition is known as ecological balance, and the system is called as 'Balanced Ecosystem'. If any disturbance occurs due to natural or manmade activities, this balance gets upset and it will be no more a balanced ecosystem. If sufficient time is allowed for restoration, a balanced ecosystem will gradually reappear, but may not resemble the original system – a new balance or equilibrium condition appears.

2.7 Components of Ecosystem

As discussed above, an ecosystem has three distinctive components that can be identified as:

- non living or abiotic component including climate regime
- living or biotic component
- source of energy – light and heat

(i) ***Abiotic Substances*****:** These comprise of inorganic and organic compounds present in the environment. The inorganic components of an ecosystem are oxygen, carbon dioxide, water, minerals etc., whereas carbohydrates, proteins, lipids, amino acids etc., are examples for organic material. The climate, light and heat can be either studied under abiotic component, or as separate entities. The predominant source of energy in the earth's biosphere is sun. The aboitic substances are circulated in the ecosystem through material cycles and energy cycles.

(ii) ***Biotic Substances*****:** Living organisms in the ecosystem – various species of plants and animals including microbes are termed as biotic components. They can be classified as producers (autotrophs) and consumers (heterotrophs).

(a) ***Producers (Autotrophs)*****:** Autotrophs produce their own food from inorganic substances, using light or chemical energy. Green plants including the unicellular algae which contain the pigment chlorophyll are producers. They take up simple substances such as water, carbon di-oxide, and oxygen, as well as inorganic nutrients and produce biological molecules needed for life from the inorganic substances. This production activity is vital for the existence of the ecosystem as the products of photosynthesis support the life on earth.

The overall effect of photosynthesis is to unite the hydrogen atoms of water with the atoms of carbon di-oxide to form carbohydrate. In the process oxygen gets released. A generalized photosynthesis reaction can be represented as:

$$H_2O + CO_2 + \text{light energy} \xrightarrow{\text{Chlorophyll}} \text{Carbohydrate} + 0_2$$

Energy obtained from solar radiation plays the key role in this process. Hence the photosynthetic activity is essentially brought about during day time, although some insignificant amount of photosynthesis takes place during night time utilizing the faint light emitted from the heavenly bodies.

(b) ***Consumers (heterotrophs)*****:** The heterotrophs do not have the ability to produce their own food. All these species are consumers. Bacteria, although belong to plant kingdom are not capable of production and are classified as consumers. The animals which feed on plants are called herbivores. They are primary consumers. Those feeding on animals are called as carnivores which are secondary consumers. Another category of consumers which feed on both plants and animals are called as omnivores.

2.8 Classification of Ecosystems

Ecosystems are broadly classified as:

Terrestrial Ecosystems - which encompass the activities that take place on land, and

Aquatic ecosystems - the system that exists in water bodies

These ecosystems can be further subdivided as:

Terrestrial ecosystem - Forest ecosystem,
Mountain ecosystem
Desert ecosystem
Grassland ecosystem
Urban ecosystem

Aquatic ecosystem - Marine ecosystem
Fresh water ecosystem
Esturine ecosystem

Engineered ecosystem - An ecosystem which is fully designed and controlled by man is called 'Engineered ecosystem'. A paddy field or a fish pond can be quoted as an example for this ecosystem.

Examples of Ecosystems: Two examples are given in the following sketches.

Fig. 2.5 Classification of Ecosystem

(a) **Components of the Terrestrial Ecosystem:** The following figure illustrates the different trophic levels for an arbitrary terrestrial ecosystem

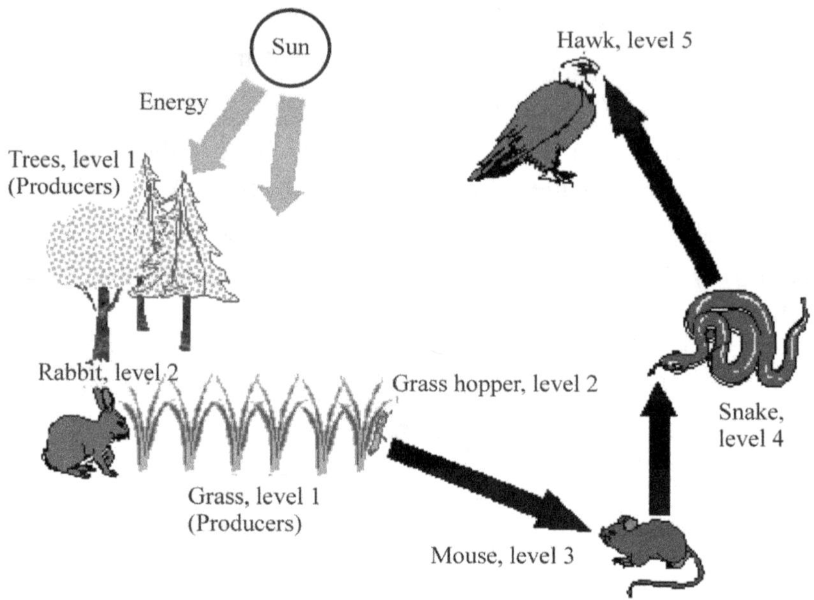

Fig. 2.6(a) Terrestrial Ecosystem

Organisms residing higher on the trophic scale have less energy available to them, as illustrated below:

There is one type of extensive terrestrial ecosystem due solely to human activities and eight types that are natural ecosystems. Those natural ecosystems reflect the variation of precipitation and temperature over Earth's surface. The smallest land areas are occupied by tundra and temperate grassland ecosystems, and the largest land area is occupied by tropical forest. The most productive ecosystems are temperate and tropical forests, and the least productive are deserts and tundras. Cultivated lands, which together with grasslands and savannas utilized for grazing are referred to as agroeco systems, are of intermediate extent and productivity. Because of both their a real extent and their high average productivity, tropical forests are the most productive of all terrestrial ecosystems, contributing 45% of total estimated net primary productivity on land.

(b) **Components of the Aquatic (pond) ecosystem:** Different components of aquatic ecosystem are shown in Fig. 2.6(b).

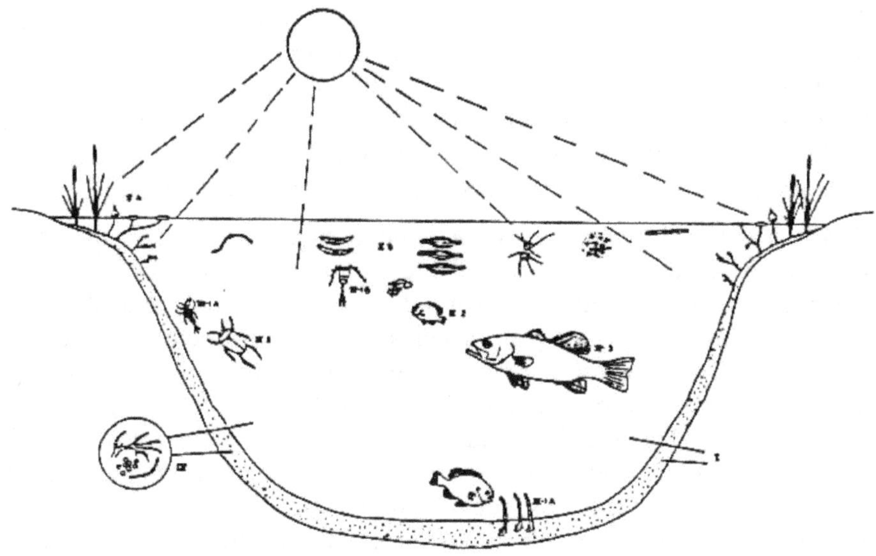

Fig. 2.6(b) Aquatic Ecosystem

Abiotic : Water, dissolved oxygen, carbon dioxide, inorganic salts such as chlorides, nitrates, phosphates of calcium, sodium, potassium etc., A large number of organic compounds such as organic acids are also present.

Biotic component : Both producers and consumers are present in the pond

Producers: In a freshwater pond, two types of producers are present - the large plants- floating or growing along the shoreline, and the floating and suspended microscopic plants. Mostly the later variety is made up of different types of algae. They are distributed throughout the water as deep as sunlight penetrates. These small plants are called as phytoplankton. Individual algae cells are not visible, but when they are present in large quantity give a greenish hue to the water body.

Consumers: Alongside the producers, a pond ecosystem contains consumers such as fish, insects, crabs etc. They include both primary consumers (herbivores) and secondary consumers (carnivores). Tertiary consumers feeding on carnivores can also be present. These consumers are visible to naked eye, and hence called as macro consumers. There are microscopic sized consumers also. They are called as Zooplanktons, and are present at the surface of the water and as well at the bottom (benthos).

The pond ecosystem accommodates a major consumer form including bacteria, fungus etc., which are called as decomposers. They are micro consumers and play a major role in breaking down the waste products of macro consumers, and dead consumer and producers organisms. But for the decomposers, the ecosystem cannot exist as it gets overloaded with waste products and dead organisms. They are great scavengers.

Algal – Bacterial Symbiosis: In a pond ecosystem bacteria, the main decomposers feed on the biodegradable organic matter available to them in the form of waste matter discharged by animal species, and the dead organisms both animal and plant species. They consume oxygen for bio chemical oxidation of the organic matter, and for their own respiration. As a consequence carbon dioxide is liberated. This carbon dioxide is taken up by the algae that is abundantly available. The growth of algae is promoted by the presence of nutrients in water. Algae being able to carry out photosynthesis in the presence of sunlight take up the CO_2 and release O_2 which is readily taken up by bacteria. This cyclic activity is called Algal-Bacterial symbiosis which keeps the pond ecosystem in a balanced condition.

2.9 Natural Cycles of the Environment

Nutrients, unlike energy are recycled in the ecosystem. There are about 40 chemical elements considered to be essential for living organisms. Materials are in limited quantity in the earth's system and to keep the system going continuously the only possibility is to regenerate the materials. The unique method evolved in nature is recycling materials continuously is by linking them in cyclic changes.

The macro-nutrients are C, H, O, P, K, I, N, S, Mg, Ca, etc., which have cycles with atmosphere while micro-nutrients like Cu, Fe, Co, etc., are soil based form edophic cycles. The bio-geo-chemical cycles are of two varieties – sedimentary cycles and gaseous cycles. In sedimentary cycles the main reservoir is the soil, the sedimentary and other types of rocks of earth's crust. The gaseous cycles have their main reservoir of nutrients in the atmosphere and oceans. Examples are the oxygen, carbon, nitrogen, sulphur, etc. Both are driven by the flow of energy and both are tied up with the water cycle or the hydrologic cycle. In nutrient cycle, various chemical compounds of the main element are transferred while in hydrologic cycle a compound i.e., water is circulated as solid liquid and vapour phase.

2.9.1 Water (Hydrologic) cycle

The hydrologic cycle involves a continuous exchange of water between sea, atmosphere, land and living animals through massive evaporation of water from the ocean, cloud formation and precipitation as outlined in Fig. 2.7.

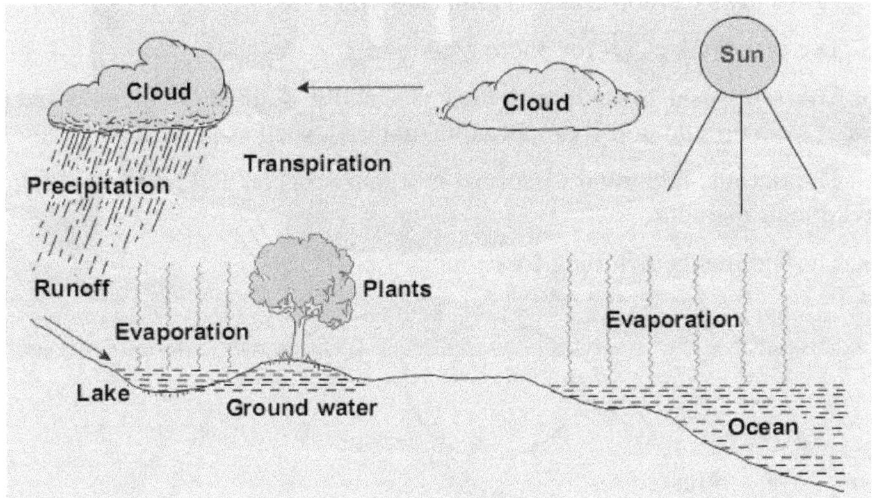

Fig. 2.7 Water Cycle

The land surface and water surfaces on earth lose water by evaporation by solar energy. Normal evaporation of water from ocean exceeds precipitation by rain into seas by 10%.

- This 10% excess which precipitates on land balances the hydrological cycle.
- Some of the precipitated rain seeps into the soil as ground water.
- Ground water moves up by capillary action and there by maintains a continuous supply of water to the surface layer of soil.

The water from the surface layer of the soil is absorbed by plants, which in turn is returned to atmosphere through transpiration.

- Surface water or runoff flows into streams, rivers, lakes and catchment areas or reservoirs.
- Animals also take water which is also returned to the atmosphere through evaporation.
- Thus there is always a balanced continuous cycling of water between earth's surfaces and atmosphere.

2.9.2 Carbon cycle

The carbon cycle is the biogeochemical cycle by which carbon is exchanged among the biosphere, pedosphere, geosphere, hydrosphere, and atmosphere of the Earth. The carbon cycle is usually thought of as four major reservoirs of carbon interconnected by pathways of exchange. Fig. 2.8 shows Carbon Cycle. These reservoirs are:

The plants take CO_2 for photosynthesis.

The terrestrial biosphere, which is usually defined to include fresh water systems and non-living organic material, such as soil carbon.

The oceans, including dissolved inorganic carbon and living and non-living marine biota,

The sediments including fossil fuels.

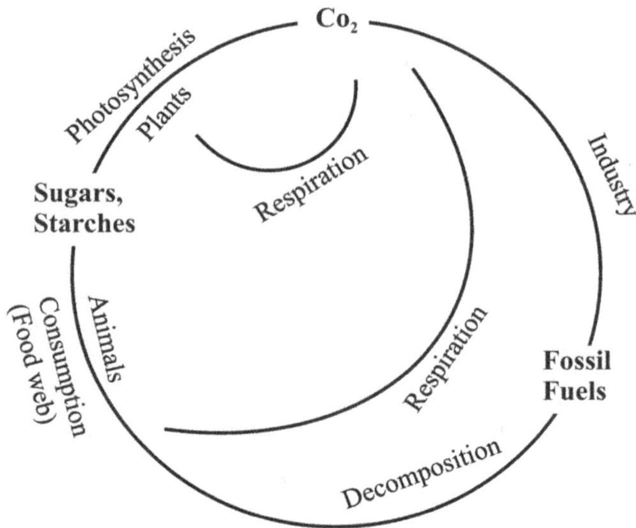

Fig. 2.8 Carbon Cycle

CO_2 in atmosphere had taken in by plants to make sugar, Primary consumers (animals) eat plants-obtain carbon, secondary consumers eat primary consumers. Carbon returns to ecosystem through respiration, decomposition.Fossil fuel carbon released by burning (human activity)

Carbon is an essential constituent of carbohydrates, proteins, fats and a large number of organic compounds. CO_2 of the atmosphere and that dissolved in the natural waters is the main source of carbon. Green plants use CO_2 in the process of photosynthesis to make carbohydrates. In doing so the green plants lock the radiant energy of the sun in the synthesized food. This energy is utilized by all living beings for their own activities. The evolved oxygen by the process of photosynthesis is used for most of the living things, the plants and animals. Thus all animals depend for their food on plants and animals. Thus all animals depend for their food on plants directly or indirectly. All organic compounds are also oxidized to CO_2 and water, both of which are utilized by plants in the process of photosynthesis.

$$C_6H_{12}O_6 + 6O_2 \rightarrow 6CO_2 + 6H_2O + \text{Free Energy}$$

The water goes down into the soil for the use of plants. The atmosphere and natural waters must be replenished with CO_2. Most of the CO_2 is returned to atmosphere and natural water by plants and animals through the process of respiration. Bacteria and fungi also return CO_2 to the atmosphere and natural water into the soil by acting chemicals upon the dead plants and animals and their waste such as urine and faeces. It should also be noted that coal, petroleum, etc., are also noted that coal, petroleum, etc., are also the part of carbon cycle and are formed in nature by living organisms. Decomposition of micro-organism are very important in breaking down dead material with the release of carbon back to the carbon cycle. All the carbon of plants, herbivores, carnivores and decomposers is not respired, but some are fermented and some are stored. The carbon compounds such as methane that are lost to the food chain after fermentation are readily oxidized to CO_2 by a number of reactions occurring in the atmosphere

2.9.3 Nitrogen cycle

Of all the elements that plants absorb from soil, nitrogen is the most important element for plants growth. It is required for amino acids, proteins, enzymes, chlorophyll, nucleic acids and many other compounds. But the atmospheric nitrogen is not utilized directly. Nitrogen undergoes many changes in the nitrogen cycle like, nitrogen fixation, nitrogen assimilation, ammonification, nitrification, denitrification (Fig. 2.9).

Nitrogen fixation or conversion of free nitrogen of atmosphere into biologically acceptable form or nitrogenous compounds is referred to as nitrogen fixation. The fixation of nitrogen requires an investment of energy. Before nitrogen can be fixed, it must be activated so that the molecular nitrogen must be split into two atoms of free nitrogen. In physico-chemical process nitrogen combines with oxygen (as ozone) during lightening or electrical discharges in the clouds and produces different oxygen oxides. These nitrogen oxides get dissolved in rain water and on earth's surface, they react with mineral compounds to form nitrates and nitrogenous compounds.

$$N_2 + 2(O) \rightarrow \text{electric changes} \rightarrow 2NO$$

$$2NO + 2(O) \rightarrow 2NO_2$$

$$2NO_2 + (O) \rightarrow N_2O_5$$

$$N_2O_5 + H_2O \rightarrow 2HNO_3$$

$$2HNO_3 + CaCO_3 \rightarrow Ca(NO_3)_2 + CO_2 + H_2O$$

Biological nitrogen fixation is carried by some blue-green algae in the oceans, lakes and soils. Symbiotic bacteria (rhizobium) living in root nodules of leguminous plants and few other plants can fix nitrogen. Certain free living nitrogen fixing bacteria also fix nitrogen. Fixed nitrogen means nitrogen incorporated in a chemical compound that can be utilized by plants and animals. The actual fixation steps involves with two atoms of nitrogen combined with 3 atoms of hydrogen to form 2 molecules of ammonia. The activation and fixing, the two steps require a net input of 147 Kilo Calories. Once ammonia or ammonium ion appeared in the soil, it can be absorbed by the roots of plants and the nitrogen can be incorporated into amino acids and then to protein.

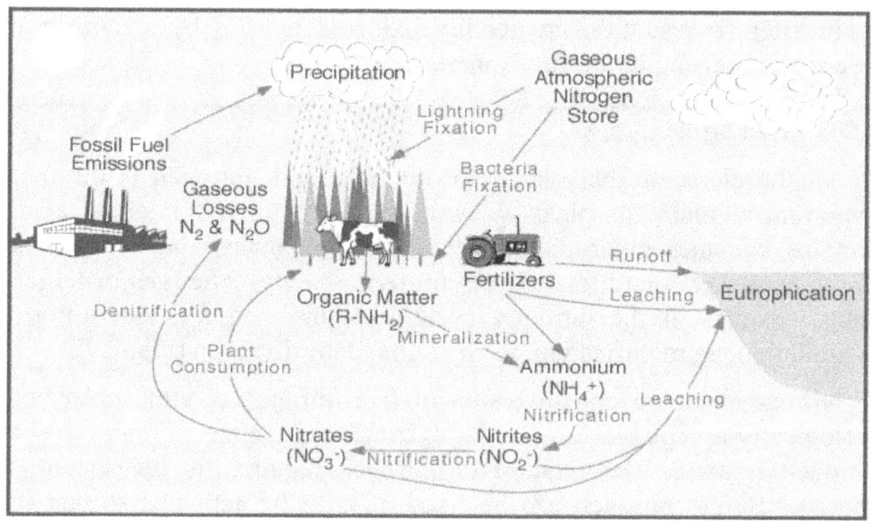

Fig. 2.9 Nitrogen Cycle

Nitrogen assimilation means that the inorganic nitrates, nitrites or ammonia must be incorporated into organic compounds. Ammonification means that the dead organic remains of plants and animals and excreta are acted upon by bacteria, actinomycetes releasing nitrogen as ammonia. Enitrification means conversion of ammonia into nitrate by nitrosomonas, nitro coccus, in oceans and soils. Conversion of nitrite into nitrate by nitrobacter is also nitrification.Denitrofication in conversion of nitrite and nitrate into nitrogen by Thiobacillus denitrifications, micro coccus, denitrificans, *pseudomonas aerusinosa*, etc.

2.9.4 Sulphur cycle

Unlike carbon and oxygen cycles (gaseous cycles), sulphur and phosphorus cycles are sedimentary cycle. Sulphur is present normally as sulphates or sulphides. In sulphur springs and volcanic eruptions sulphur di-oxide is present to some extent. Sulphur is a component of 3 amino acids. Sulphur cycle is going to be important from protein synthesis point of view. Almost all proteins contain these amino acids. Sulphur is also present in the fossil fuels which emit sulphur di-oxide, in the automobile exhaust. Under anaerobic conditions sulphates are used to supply oxygen for sulphur organisms. In some of the sulphur bacteria elements of sulphur is precipitated. Hydrogen sulphide produced under anaerobic conditions can be oxidized to suplhur or sulphates. Sulphur di-oxide in the atmosphere gets converted to sulphorous and sulphuric acid causing the acid rain problem in many urban and industrial areas. In sewers, because of anaerobic conditions, H_2S is produced. This get oxidized with oxygen present in the sewer pipe and become SO_2 which dissolves in water to form sulphuric acid. Accumulation of this inside the pipe results in 'crown corrosion' in sewers.

Sulphur cycle links soil, water (Fig. 2.10). Sulphur also occurs in soils and rocks as sulphides (FeS, ZnS, etc.). Except a few organisms which need organic form of suplhur as amino acids and cystein, most of the organisms take sulphur as inorganic sulphates. Under aerobic conditions sulphur can be reduced to directly sulphides.

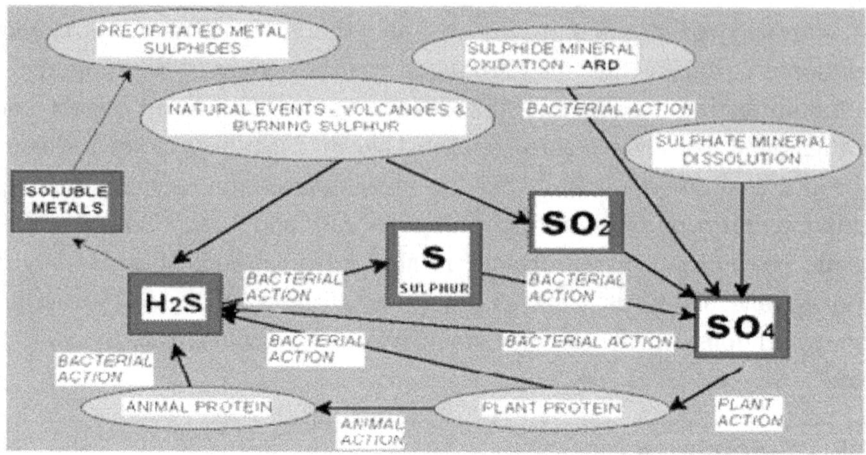

Fig. 2.10 Sulphur Cycle

Green and purple photosynthetic bacteria use hydrogen of H_2S as oxygen acceptor in reducing carbon-di-oxide. Green bacteria are also to oxidize sulphide to elemental sulphur, whereas purple sulphur bacteria can carry oxidation of sulphate stage. In the ecosystem sulphur is transferred from autotrophs to animals, their to decomposers and finally it returns to environment through death and decay of dead organisms.

Sedimentary nature of sulphur cycling involves precipitation of sulphur in presence of iron, under anaerobic conditions. Sulphides of iron, copper, zinc, cadmium, cobalt are insoluble in neutral and alkalie waters and sulphur is bound to limit the amount of these elements.

2.10 Biodiversity

Biodiversity includes all organisms, species, and populations; the genetic variation among these; and all their complex assemblages of communities and ecosystems. It also refers to the interrelatedness of genes, species, and ecosystems and their interactions with the environment. Usually three levels of biodiversity are discussed—genetic, species, and ecosystem diversity.

- *Genetic diversity* is all the different genes contained in all individual plants, animals, fungi, and microorganisms. It occurs within a species as well as between species.
- *Species diversity* is all the differences within and between populations of species, as well as between different species.

- *Ecosystem diversity* is all the different habitats, biological communities, and ecological processes, as well as variation within individual ecosystems.

2.10.1 Biodiversity Threats

The loss of biodiversity is a significant issue for scientists and policy-makers and the topic is finding its way into living rooms and classrooms. Species are becoming extinct at the fastest rate known in geological history and most of these extinctions have been tied to human activity. The main threats to Earth's biodiversity are (i) habitat loss and degradation, (ii) Overharvesting, (ii) non-native species, (iv) global environmental change, (v) pollution and (vi) Alterations in ecosystem composition.

(i) *Habitat Loss and Degradation:* Habitat loss is the biggest threat to the world's biodiversity. The causes of habitat loss are numerous, including human population growth and economic development. Existing habitats are altered by agricultural conversion, erosion, deforestation and urbanization. Habitat loss is a particularly serious problem in areas where population growth, poverty, civil war, and land ownership systems cause people to use resources in unsustainable ways. Habitat loss and destruction, usually as a direct result of human activity and population growth, is a major force in the loss of species, populations, and ecosystems.

(ii) *Over harvesting:* The overuse and overharvesting of species haveled to a drastic decline in once-abundant fish populations and to near-extinction of species such as the rhinoceros. Methods of producing goods more efficiently and effectively, such as tropical tree plantations and single-crop farms, have been developed to meet the increasing need for both food and fuel. However, these methods can be detrimental to biodiversity. The over-exploitation (over-hunting, over-fishing, or over-collecting) of a species or population can lead to its demise.

(iii) *Non-Native Species*: Accidentally or deliberately introducing non-native species into habitats is another threat to biodiversity. These species can come to dominate a habitat, driving native species to extinction and changing the way the ecosystem operates. For example, in Africa the introduction of the Nile erch into Lake Victoria resulted in two-thirds of the lake's native cichlid (fish) species (about 200) going extinct .In the United States, a fungus that arrived from Asia in the early 1900s nearly wiped out the American chestnut tree. Seven moth species that fed only on these chestnuts may also now be extinct. The introduction of exotic (non-native) species can disrupt entire ecosystems and impact populations of native plants or animals. These invaders can adversely affect native species by eating them, infecting them, competing with them, or mating with them

(iv) *Global Environmental Change*: Large-scale environmental changes such as climate change, ozone depletion, tropical deforestation, and acid rain can also threaten biodiversity. One example is the ocean warming off the coast of San Diego, California. Zooplankton, fish, and seabird populations have declined by 80%, reducing the biological richness of the area. Global climate change can alter environmental conditions. Species and populations may be lost if they are unable to adapt to new conditions or relocate

(v) *Pollution:* Human-generated pollution and contamination can affect all levels of biodiversity. Pollution within habitats can severely affect biodiversity. Each year, the pesticide poisoning of honeybees costs American farmers millions of dollars. Certain other chemical pollutants are suspected of disrupting the endocrine systems of amphibians, reducing their reproductive rate and their availability as food for other species.

(vi) *Alterations in ecosystem composition*: Loss or decline of a species, can lead to a loss of biodiversity. For example, efforts to eliminate coyotes in the canyons of southern California are linked to decreases in song bird populations in the area. As coyote populations were reduced, the populations of their prey, primarily raccoons, increased. Since raccoons eat bird eggs, fewer coyotes led to more raccoons eating more eggs, resulting in fewer song birds.

2.10.2 Importance of Biodiversity

The diversity of life enriches the quality of our lives in ways that are not easy to quantify. Biodiversity is intrinsically valuable and is important for our emotional, psychological, and spiritual well-being. Some consider that it is an important human responsibility to be stewards for the rest of the world's living organisms.

Diversity breeds diversity. Having a diverse array of living organisms allows other organisms to take advantage of the resources provided. For example, trees provide habitat and nutrients for birds, insects, other plants and animals, fungi, and microbes.

Humans have always depended on the Earth's biodiversity for food, shelter, and health. Biological resources that provide goods for human use include:

- Food-species that are hunted, fished, and gathered, as well as those cultivated for agriculture, forestry, and aquaculture;
- Shelter and warmth-timber and other forest products and fibers such as wool and cotton;
- Medicines-both traditional medicines and those synthesized from biological resources and processes.

Biodiversity also supplies indirect services to humans which are often taken for granted.

These include drinkable water, clean air, and fertile soils. The loss of populations, species, or groups of species from an ecosystem can upset its normal function and disrupt these ecological services. Recent declines in honeybee populations may result in a loss of pollination services for fruit crops and flowers. Biodiversity provides medical models for research into solving human health problems. For example, researchers are looking at how seals, whales, and penguins use oxygen during deep-water dives for clues to treat people who suffer strokes, shock, and lung disease.

The Earth's biodiversity contributes to the productivity of natural and agricultural systems. Insects, bats, birds, and other animals serve as pollinators. Parasites and predators can act as natural pest controls. Various organisms are responsible for recycling organic materials and maintaining the productivity of soil.

Genetic diversity is also important in terms of evolution. The loss of individuals, populations, and species decreases the variety of genes-the material needed for species and populations to adapt to changing conditions or for new species to evolve.

2.10.3 Conservation of Biodiversity

Having been aware of the value of the biodiversity and the consequence of loss of biodiversity, it is very obvious for the need for conservation. The main objectives and benefits of biodiversity conservation are as follows.

- Conservation of biological diversity leads to conservation of ecological diversity so that the continuity of food chain can be preserved.
- The genetic diversity of plant and animals are preserved.

- It ensures sustainable utilization of life support system on earth.
- It provides a vast knowledge of potential use to the scientific community.
- By preserving the genes of wild animals and plants through gene bank we can make use of them in future.
- The existence of biodiversity provides benefit to the society by way tourism and recreation.

(i) In-situ conservation

Conserving the animals and plants in their own natural habitats is known as In-situ conservation. This includes the establishment of, National parks, Sanctuaries, Biosphere reserves, Nature reserves, Reserved forests, Preservation plots.

(ii) Ex-situ conservation

The biodiversity conservation being carried out at the outside of the natural areas is known as Ex-situ (off-site) conservation. Animals and Plants are reared or cultivated in areas like zoological and botanical parks. Reintroduction of animals or plants into habitat from where it has become extinct is another form of Ex-situ conservation. For example, the crocodile species, Gangetic gharials has been re-introduced in the rivers of Uttar Pradesh, Madhya Pradesh and Rajasthan where it has become extinct. Seed bank, gene bank, botanical, horticultural, and recreational gardens are some of the important centers for Ex-situ conservation in India.

2.11 'Food Chain' and 'Food Web'

Energy is passed from one organism to another in a complex network like a spider's web. A food chain is the sequence of who eats whom in a biological community (an ecosystem) to obtain nutrition. The transfer of food and energy from the plant source through a series of organisms that consume and are consumed is called food chain.

(a) ***Components of Food Chain*****:** The food chain consists of four main parts.

(i) *The Sun*: which provides the energy for everything on the planet

(ii) *Producers:* these include all green plants. These are also known as autotrophs, since they make their own food. Producers are able to harness the energy of the sun to make food. Ultimately, every (aerobic) organism is dependent on plants for oxygen (which is the waste product from photosynthesis) and food (which is produced in the form of glucose through photosynthesis). They make up the bulk of the food chain or web.

(iii) *Consumers:* In short, consumers are every organism that eats something else. They include herbivores (animals that eat plants), carnivores (animals that eat other animals), parasites (animals that live off of other organisms by harming it), and scavengers (animals that eat dead animal carcasses). Primary consumers are the herbivores, and are the second largest biomass in an ecosystem. The animals that eat the herbivores (carnivores) make up the third largest biomass, and are also known as secondary consumers. This continues with tertiary consumers, etc.

(iv) *Decomposers:* These are mainly bacteria and fungi that convert dead matter into gases such as carbon and nitrogen to be released back into the air, soil, or water. Fungi and other organisms that break down dead organic matter are known as saprophytes. Even though most of us hate those mushrooms or molds, they actually play a very important role. Without decomposers, the earth would be covered in trash. Decomposers are necessary since they recycle the nutrients to be used again by producers.

(b) ***Trophic Levels***: The different species in a food chain are called trophic levels. Each food chain has three major trophic levels- Producers, consumers and decomposers. The trophic level of an organism is the position it holds in a food chain. Figure shows the relational biomass of each of the major groups in the food chain:

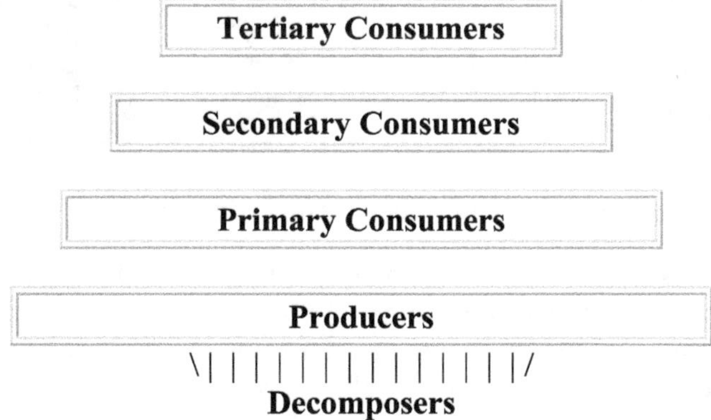

Fig. 2.11 Relational biomass

1. Primary producers (organisms that make their own food from sunlight and/or chemical energy from deep sea vents) are the base of every food chain - these organisms are called autotrophs.
2. Primary consumers are animals that eat primary producers; they are also called herbivores (plant-eaters).
3. Secondary consumers eat primary consumers. They are carnivores (meat-eaters) and omnivores (animals that eat both animals and plants).
4. Tertiary consumers eat secondary consumers.
5. Quaternary consumers eat tertiary consumers.

Food chains "end" with top predators, animals that have little or no natural enemies. Fig. 2.12 shows a Sample food Chain in 3 biomes.

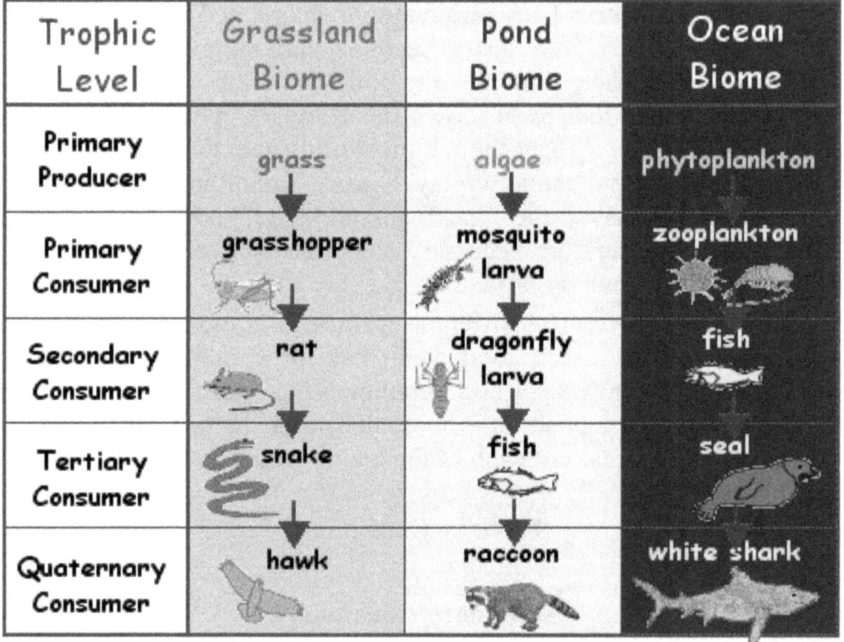

Trophic Level	Grassland Biome	Pond Biome	Ocean Biome
Primary Producer	grass	algae	phytoplankton
Primary Consumer	grasshopper	mosquito larva	zooplankton
Secondary Consumer	rat	dragonfly larva	fish
Tertiary Consumer	snake	fish	seal
Quaternary Consumer	hawk	raccoon	white shark

Fig. 2.12 Sample food Chain in three biomes

(c) ***Food Web***: The arrows in a food chain show the flow of energy, from the sun to a top predator. As the energy flows from organism to organism, energy is lost at each step. A network of many food chains is called a food web. Various food chains are often interlinked at different trophic levels to form a complex interaction between different species from the point of view of food. This network is called the food web (Fig. 2.13).

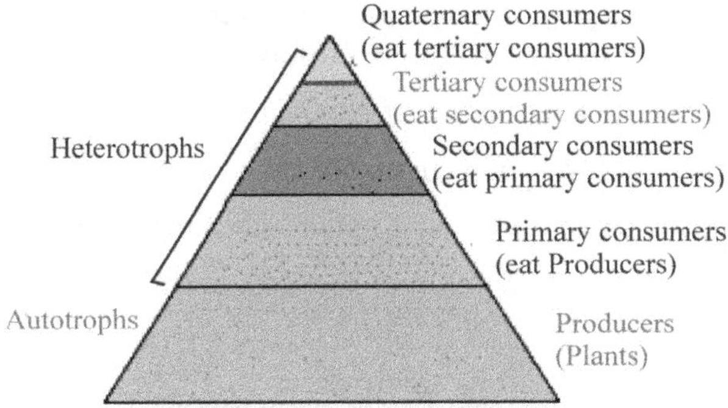

Fig. 2.13 shows a sketch of food web

Review Questions

1. Explain the components, characteristics and biodiversity of Forest ecosystem.
2. Explain the structure and functional features of Aquatic ecosystem.
3. Explain: Ecosystem, Advantages of Ecosystem, Energy flow in Ecosystem, food chain, food webs, Ecological pyramid and Ecological succession
4. Discuss about the prime characteristics of
5. Forest Ecosystem
6. Grassland Ecosystem
7. Desert Ecosystem
8. Aquatic Ecosystem
9. What is biodiversity? Explain genetic diversity, species biodiversity and ecosystem biodiversity
10. Explain biosphere
11. Explain Carbon and Nitrogen cycles.
12. Explain the Sulphur and Phosphorus cycles.
13. Describe the types, features, structures and function of various components of an Ecosystem. (Forest, Grassland, Pond and aquatic Ecosystem).

14. Explain In-situ and Ex-situ conservation of biodiversity.(or) What are the measures recommended for conservation of biodiversity.
15. Explain the various threats to Biodiversity and the measures recommended for conservation of Biodiversity.
16. Explain in detail the structure of Atmosphere.
17. Write short notes on Producers, Consumers and Decomposers.
18. What are Biochemical cycles? Explain their importance?
19. Define Eco-System. What are the structural components of an Ecosystem?
20. What is biodiversity and its significance

CHAPTER 3

Air Pollution

Air Pollution: Air pollutants, Classification, (Primary & Secondary Pollutants) Adverse effects of pollutants. Causes of Air pollution chemical, photochemical, Greenhouse Effect, Ozone layer depletion, Acid rain.

3.1 Pollution

The word Pollution has been derived from the Latin word 'Pollutionem' (meaning to defile or make dirty). Pollutant is a substance, the presence of which causes pollution. The pollutants reach us through the air we breathe, the water we drink, the food we eat and the sound we hear.

Pollution is defined as "an undesirable change in the physical, chemical or biological characteristics of our air, land and water that will harmfully affect human life or that of desirable species, living conditions etc."

There are seven main types of Pollutions in the environment (i) Air Pollution (ii) Water pollution (iii) Land or Soil Pollution (iv) Noise Pollution (v) Radiation Pollution.

3.2 Air Pollution

Air pollution is the presence of contaminants in atmosphere in quantities such that it is injurious to human, plant animal life and property. An air pollutant is known as a substance in the air that can cause harm to humans and the environment. Pollutants can be in the form of solid particles, liquid droplets, or gases. In addition, they may be natural or man-made. Air pollution is the introduction of chemicals, particulate matter, or biological materials that cause harm or discomfort to humans or other living organisms, or damages the natural environment into the atmosphere.

(A) Primary Pollutant

Pollutants can be classified as either **primary or secondary**. A primary pollutant is an air pollutant emitted directly from a source. Usually, primary pollutants are substances directly emitted from a process, such as ash from a volcanic eruption, the carbon monoxide gas from a motor vehicle exhaust or sulfur dioxide released from factories.

Note that some pollutants may be both primary and secondary: that is, they are both emitted directly and formed from other primary pollutants.

Major Primary Pollutants Produced by Human Activity Include

***Sulfur oxides* (*SOx*)**: especially sulfur dioxide, a chemical compound with the formula SO2. SO2 is produced by volcanoes and in various industrial processes. Since coal and petroleum often contain sulfur compounds, their combustion generates sulfur dioxide. Further oxidation of SO2, usually in the presence of a catalyst such as NO2, forms H2SO4, and thus acid rain. This is one of the causes for concern over the environmental impact of the use of these fuels as power sources.

***Nitrogen oxides* (*NOx*):** especially nitrogen dioxide are emitted from high temperature combustion. Can be seen as the brown haze dome above or plume downwind of cities. Nitrogen dioxide is the chemical compound with the formula NO2. It is one of the several nitrogen oxides. This reddish-brown toxic gas has a characteristic sharp, biting odor. NO2 is one of the most prominent air pollutants.

Carbon monoxide: is a colourless, odourless, non-irritating but very poisonous gas. It is a product by incomplete combustion of fuel such as natural gas, coal or wood. Vehicular exhaust is a major source of carbon monoxide.

Carbon dioxide (CO_2): a greenhouse gas emitted from combustion but is also a gas vital to living organisms. It is a natural gas in the atmosphere.

Volatile organic compounds: VOCs are an important outdoor air pollutant. In this field they are often divided into the separate categories of methane (CH4) and non-methane (NMVOCs). Methane is an extremely efficient greenhouse gas which contributes to enhanced global warming. Other hydrocarbon VOCs are also significant greenhouse gases via their role in creating ozone and in prolonging the life of methane in the atmosphere, although the effect varies depending on local air quality. Within the NMVOCs, the aromatic compounds benzene, toluene and xylene are suspected carcinogens and may lead to leukemia through prolonged exposure. 1, 3-butadiene is another dangerous compound which is often associated with industrial uses.

Particulate matter: Particulates, alternatively referred to as particulate matter (PM) or fine particles, are tiny particles of solid or liquid suspended in a gas. In contrast, aerosol refers to particles and the gas together. Sources of particulate matter can be man made or natural. Some particulates occur naturally, originating from volcanoes, dust storms, forest and grassland fires, living vegetation, and sea spray. Human activities, such as the burning of fossil fuels in vehicles, power plants and various industrial processes also generate significant amounts of aerosols. Averaged over the globe, anthropogenic aerosols-those made by human activities-currently account for about 10 percent of the total amount of aerosols in our atmosphere. Increased levels of fine particles in the air are linked to health hazards such as heart disease, altered lung function and lung cancer.

Persistent free radicals connected to airborne fine particles could cause cardiopulmonary disease.

Toxic metals, such as lead, cadmium and copper.

***Chlorofluorocarbons* (*CFCs*):** harmful to the ozone layer emitted from products currently banned from use.

***Ammonia* (*NH3*):** emitted from agricultural processes. Ammonia is a compound with the formula NH3. It is normally encountered as a gas with a characteristic pungent odor. Ammonia contributes significantly to the nutritional needs of terrestrial organisms by serving as a precursor to foodstuffs and fertilizers. Ammonia, either directly or indirectly, is also a building block for the synthesis of many pharmaceuticals. Although in wide use, ammonia is both caustic and hazardous.

Odors: such as from garbage, sewage, and industrial processes.

Radioactive pollutants: produced by nuclear explosions, war explosives, and natural processes such as the radioactive decay of radon.

(B) **Secondary pollutants** are not emitted directly. Rather, they form in the air when primary pollutants react or interact. An important example of a secondary pollutant is ground level ozone one of the many secondary pollutants that make up photochemical smog. A secondary pollutant is not directly emitted as such, but forms when other pollutants (primary pollutants) react in the atmosphere.

Examples of a secondary pollutant include ozone, which is formed when hydrocarbons (HC) and nitrogen oxides (NOx) combine in the presence of sunlight; NO2, which is formed as NO combines with oxygen in the air; and acid rain, which is formed when sulfur dioxide or nitrogen oxides react with water.

Secondary Pollutants Include

Particulate matter formed from gaseous primary pollutants and compounds in photochemical smog. Smog is a kind of air pollution; the word "smog" is a portmanteau of smoke and fog. Classic smog results from large amounts of coal burning in an area caused by a mixture of smoke and sulfur dioxide. Modern smog does not usually come from coal but from vehicular and industrial emissions that are acted on in the atmosphere by sunlight to form secondary pollutants that also combine with the primary emissions to form photochemical smog.

***Ground level ozone** (O_3)* formed from NOx and VOCs. Ozone (O3) is a key constituent of the troposphere (it is also an important constituent of certain regions of the stratosphere commonly known as the Ozone layer). Photochemical and chemical reactions involving it drive many of the chemical processes that occur in the atmosphere by day and by night. At abnormally high concentrations brought about by human activities (largely the combustion of fossil fuel), it is a pollutant, and a constituent of smog.

Peroxyacetyl nitrate (PAN): similarly formed from NOx and VOCs.

A large number of minor hazardous air pollutants. Persistent organic pollutants (POPs) are organic compounds that are resistant to environmental degradation through chemical, biological, and photolytic processes. Because of this, they have been observed to persist in the environment, to be capable of long-range transport, bioaccumulate in human and animal tissue, biomagnify in food chains, and to have potential significant impacts on human health and the environment.The main pollutants in the atmosphere are SO_2 (sulphur dioxide), CO (carbon monoxide), oxides of nitrogen, particulate matter and lead.

3.2.1 Sulphur Dioxide

Sources: The main sources of SO_2 are given as.

- Combustion of fossil fuels-coal and crude oil contain up to 3% sulphur.
- Roasting of ores sulphide ores on roasting are converted to sulphur trioxide. This, when let into the atmosphere, combines with the moisture in the atmosphere to form sulphuric acid.

  For example, roasting of galena, the sulphide ore of lead.

$$2PbS + 3O_2 \rightarrow 2PbO + 2SO_2$$

$$2SO_2 + O_2 \rightarrow 2SO_3$$

$$H_2O + SO_3 \rightarrow H_2SO_4$$

- Oxidation of 1 H_2S – Hydrogen sulphide is formed during the decay of plants. This, on oxidation releases sulphur dioxide into the atmosphere.

$$2H_2S + 3O_2 \rightarrow 2H_2O + 2\ SO_2$$

- Volcanic eruptions also emit sulphur dioxide.

Effects of SO_2:

- Sulphur dioxide pollution in the atmosphere affects causes the following damages
- In humans : it causes
- eye irritation, cough, lung diseases including lung cancer and asthma
- In plants: it causes damage of leaves, bleaching of chlorophyll which turns leaves brown, damage to crops and to growth of plants.
- Others: Yellowing of paper and wearing away of leather are other ill effects.

Control:

- The gases evolved during combustion of fossil fuels are passed through calcium carbonate when SO_2 is converted to calcium sulphite.

$$CaCO_3 + SO_2 \rightarrow CaSO_3 + CO_2$$

- lime is added to coal and roasted at high temperature so that CaO formed combines with SO_2 to form calcium sulphate.

$$CaO + SO_2 + \frac{1}{2}\ O_2 \rightarrow CaSO_4$$

3.2.2 Carbon Monoxide

Sources:

- Oxidation of methane: Methane is formed during decay of vegetable matter. Oxidation of methane releases carbon monoxide into the atmosphere.

- Automobile exhaust- carbon monoxide is formed during the combustion of fuel such as petrol and is released into the atmosphere through the exhaust.
- Incomplete combustion of fossil fuels: coal when undergoes incomplete oxidation, forms carbon monoxide and pollutes the atmosphere.

$$2C + O_2 \rightarrow 2CO$$

- Industries: carbon monoxide is released by industries such as iron and steel and petroleum.

$$CO_2 + C \rightarrow 2CO$$

$$2CO_2 \rightarrow 2CO + O_2$$

Effects of CO:

- Haemoglobin in blood can form a complex with oxygen and hence functions as carrier of oxygen.
- When the atmosphere is polluted with carbon monoxide, on inhalation, CO combines with the hemoglobin to form carboxy haemoglobin and hence oxygen carrying capacity of the blood decreases.
- This causes, headache, dizziness, unconsciousness.
- When inhaled for a long duration it may cause even death.

Control: Using catalytic converter in automobiles.

3.2.3 Oxides of Nitrogen

Nitric oxide, nitrogen dioxide and nitrous oxide are the three main oxides of nitrogen found in the atmosphere.

Sources: The sources for the oxides of nitrogen are:

- Bacterial decomposition of nitrogenous compounds – bacteria in the soil act on the ammonium compounds present in the soil, convert them to ammonia and finally release oxides of nitrogen into the atmosphere.

$$4NH_3 + 5O_2 \rightarrow 4NO + 6H_2O$$

- Combustion during lightning – during lightning, oxygen and nitrogen in the atmosphere combine to give oxides of nitrogen.

$$N_2 + O_2 \rightarrow 2NO$$

$$2NO + O_2 \rightarrow 2NO_2$$

- Industries and automobile exhaust-Air is sucked into the IC engines. At high temperatures, nitrogen and oxygen in the air combine to form nitric oxide.

$$N_2 + O_2 \rightarrow 2NO$$

- Nitric oxide escapes through the exhaust. It gets cooled rapidly and combines with oxygen in the air to give nitrogen dioxide.

$$2NO + O_2 \rightarrow 2NO_2$$

Effects of oxides of nitrogen:

- Pollution due to oxides of nitrogen affects human and plant life.
- The oxides of nitrogen combine with moisture in the atmosphere to form nitrous and nitric acid. This leads to increase in the acidity of rain water.
- Formation of photochemical smog: oxides of nitrogen combine with hydrocarbons present in the atmosphere forming peroxyacyl nitrate.
- Peoxyacyl nitrate causes injury to plants and in human beings it causes fatigue and infection of the lungs.
- Peroxyacyl nitrate formation leads to smog (fog + smoke). Smog reduces visibility.
- Fading of dyes is caused in textiles.

Control: Using catalytic converter in automobiles. Catalytic converters use Pt/ Rh catalyst. In the presence of the catalysts, the oxides of nitrogen are converted to nitrogen and oxygen.

$$2NO_x \rightarrow N_2 + x\,O_2$$

3.2.4 Particulate Matter

Particulate matters are solid or liquid suspensions in air. They are also called aerosols. These comprise of dust particles, ash, smoke, fumes and mist.

Sources:

- Volcanic eruptions.
- Soil erosion: wind blows away soil and the dust particles are introduced into the atmosphere.
- Industrial operations such as crushing of solid materials- solid materials are crushed, ground and powdered in industries. During these operations dust is released into the atmosphere.
- Burning of coal: The noncombustible matter in coal is left behind as ash during the combustion of coal.
- Incomplete combustion of compounds containing carbon, processing of coal, cement asbestos: These operations also release dust into the atmosphere.
- Mist – condensation of vapours, sprays etc lead to dispersion of liquids in the atmosphere thus forming mist.

Effects of particulate matter: Presence of particulate matter in the atmosphere has the following effects.

- Decrease in visibility: Particulate matter interfere inn the transmission of light and hence affect visibility.
- Particulate matters enter the lungs causing wheezing, bronchitis, and asthma in human beings.
- In plants the particulate matter settle on the leaves blocking the stomata thereby affecting the plant growth.

Control: Particulate matter in the atmosphere can be controlled using

- Gravitational settling chambers
- Centrifugal separators
- Fabric filters
- Wet scrubbers
- Electrostatic or Cottrell separators

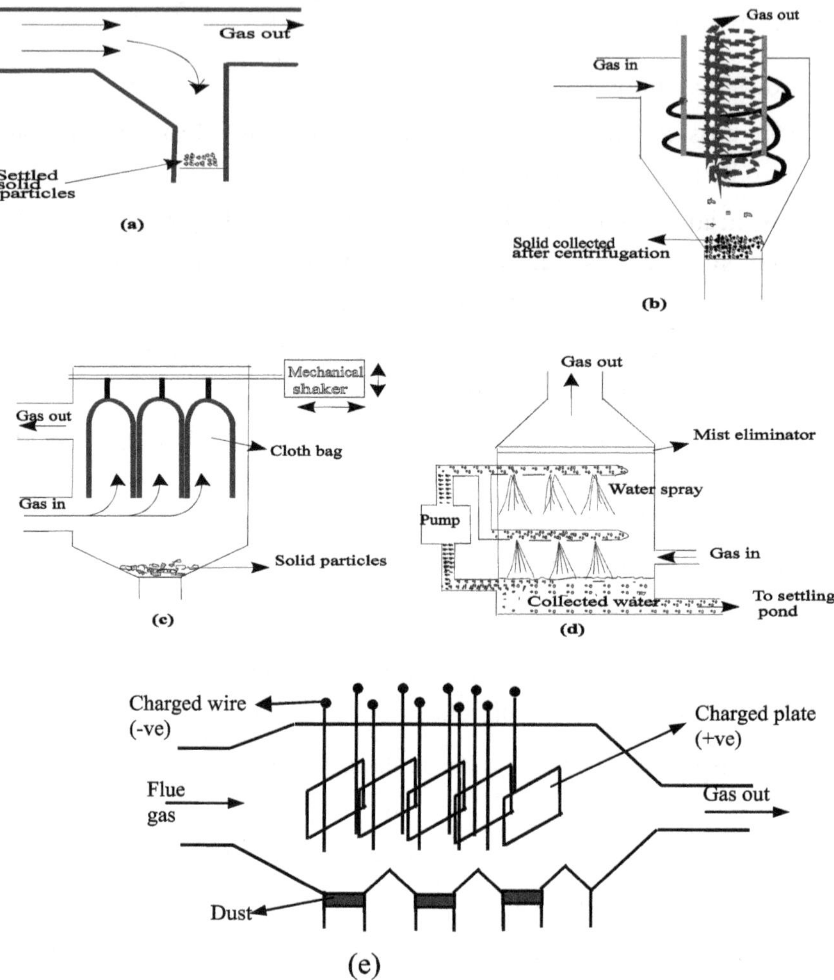

Fig. 3.1 Controlling of particulate matter in the atmosphere

(a) *Gravitational settling chambers:* Here the flue gas is allowed into a rectangular settling tank at a slow rate so that the suspended particles in the gas get deposited. The particles are later removed (**Fig. 3.1**(a)).

(b) *Centrifugal separators:* With the help of a cyclone, the gas is led into a chamber tangential to the cross section of the chamber. The gas moves in a spiral manner. Due to the centrifugal forces, the particles in the gas move towards the wall of the chamber and get deposited (**Fig. 3.1** (b)).

(c) *Fabric filters:* These consist of bags made of cotton, wool or artificial fibers ceramics. Theses can filter fine particulate matter. Flue gas is passed through a chamber containing a series of such bags. The particles are filtered and clean gas escapes. The particulates collect at the bottom and are removed periodically (**Fig. 3.1** (c)).

(d) *Wet scrubbers:* Flue gas is let into a chamber which has two sections – converging section and diverging section. The flue gas enters the converging section and water is sprayed from the top at right angles. The droplets of water take away the particulate matter in the gas (**Fig. 3.1** (d)).

(e) *Electrostatic or Cottrell separators:* The flue gas is passed into a chamber containing a series of charged plates. Between the plates wires charged to about 40000 volts are placed. As the flue gas passes through, the particles in it collide with the ionized gas molecules and the particles get charged. The positively charged particles now move towards the wire and get deposited. The negatively charged particles move towards the plates and settle. The gas which is now devoid of particulate matter goes out (**Fig. 3.1** (e)).

3.2.5 Lead pollutant

Sources:

- The exhaust from automobiles which use lead tetraethyl as anti knocking agent. When TEL is used as anti knocking agent, lead is converted to halide and released into the atmosphere. This leads to increase in the concentration of lead in the atmosphere.
- Paint pigments: Litharge and red lead (oxides of lead) and lead chromate are used as pigments. These cause lead pollution.
- Plumbing systems- lead pipes are used for plumbing and these may cause lead pollution

Effects of lead pollutant:

- Lead competes with calcium and enters the blood and bone marrow.
- The lead interferes in the manufacture of red blood corpuscles and abnormal multiplication of blood cells and thus leads to anaemia and blood cancer in human beings.

- Lead enters the blood and various organs of the body including the brain and the kidneys leading to dysfunction of the kidney and damage to the brain.

3.2.6 Photochemical Smog

Smog is a mixture of smoke and fog. Oxides of nitrogen and hydrocarbons are let into the atmosphere from automobile exhaust. The action of sunlight on these pollutants leads to the formation of peroxyacyl nitrate which causes photochemical smog.

$$N_2 + O_2 \rightarrow 2NO$$

$$2NO + O_2 \rightarrow 2NO_2$$

$$NO_2 \xrightarrow{\text{sunlight}} NO + O$$

$$O + O_2 \rightarrow O_3$$

$$RCH = CHR + O_2 \rightarrow RCO_3^* + RCH_2^*$$

(hydrocarbon)

$$RCH_2^* + O_2 \rightarrow RCH_2O_2^*$$

$$RCH_2O_2^* + NO \rightarrow NO_2 + RCH_2O^*$$

$$RCH_2O^* + O_2 \rightarrow HO_2^* + RCHO$$

$$HO_2^* + NO \rightarrow HO^* + NO_2$$

$$RCHO + HO^* \rightarrow RC \quad O^* + H_2O$$

$$RC + O^* + O_2 \rightarrow RCO_3^* \xrightarrow{NO} RC = O$$

$$|$$

$$O - O - NO_2$$

Peroxyacyl nitrate (PAN)

3.3 Green House Effect, Global Warming and Climate Change

Carbon di-oxide is a natural constituent of atmosphere, but now, its concentration is increasing at an alarming rate. According to an estimate, CO_2 level is expected to be doubled by 2030 A.D.The term ‘Green House Effect’ is also called as ‘Atmospheric Effect’, ‘Global Warming’ or ‘CO_2 Problem’.

Human activities are changing the composition as well as behavior at an unprecedented rate. The pollutants form a wide range of human activities are increasing the global atmospheric concentration of certain heat trapping gases, which act like a blanket, trapping close to the surface that would otherwise escape through the atmosphere to the outer space. This process is known as 'Green House Effect'. Green House is that body which allows the short wave length incoming solar radiation to come in, but does not allow the long wave outgoing terrestrial infra red radiation to escape. The progressive warming up of the earth's surface due to blanketing effect of manmade CO_2 in the atmosphere is called 'Green House Effect'. (Fig. 3.2).

Under normal concentrations of CO_2, the temperature of the earth's surface is maintained by the energy balance of the sun's rays that strike the planet and the heat is radiated back into the outer space. However, when concentration of CO_2 in the atmosphere increases, the thick envelope of this gas prevents the heat from being re-radiated out. The heated earth can radiate this absorbed energy as the radiation of longer wave length. A huge amount of CO_2 gets introduced into the environment from furnaces of power plants, fossil fuel burning, vehicular exhaust and breathing of animals, but the ocean may not be able to absorb this increased CO_2 and the plants also cannot utilize the whole during photosynthesis. So, much of CO_2 is still left in the atmosphere, which is supposed to be responsible for increasing the atmospheric temperature.

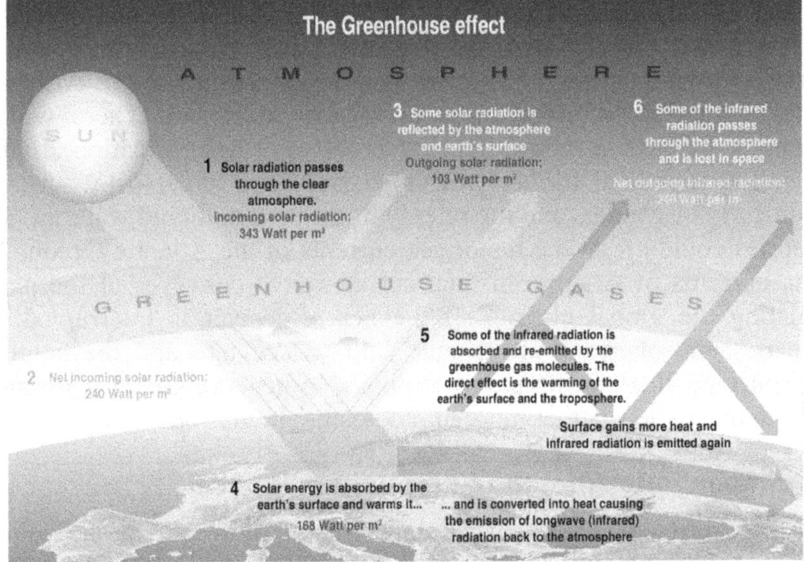

Fig. 3.2 Greenhouse effect

3.3.1 Sources

A number of industrial as well as agricultural operations generate and emit waste gases into the atmosphere. Burning of fossil fuel emit CO_2, growing paddy, or live stock releases methane. The use of aerosols and coolants in refrigerators and air conditioning devices or sprays releases chlorofluorocarbons into the atmosphere. These gases create a canopy in the atmosphere and trap the solar radiation reflected back from the earth's surface leading to atmospheric and climatic changes. The four major green house gases, which cause adverse effects are CO_2, CH_4, N_2O and CFC's. Among these CO_2 is the most common and important green house gas. In addition, ozone and SO_2 are also act as serious pollutants in causing global warming.

3.3.2 Harmful Effects of Greenhouse Effect

Some potential effects associated with the enhanced green house effect and the associated global warming is as follows.

3.4 Changes in Global Climate

Scientists believe, the average global temperature will be higher than ever in the past thousand years. The global warming trend can cause significant climatic changes. As a result of rise of temperature of earth, the oceans get warm up and sea level would rise flooding low lying regions. A slight increase in sea level could have profound effects on habitation and coastal land. In temperate regions, the winter will be shorter and warmer and the summer will be longer and hotter. A warmer climate is likely to make some cities extremely hot. There will be enormous increase in rainfall, but the problem of desertification, drought and soil erosion will further worsen. It is even postulated by scientists that melting of glaciers and the release of the resultant cold water in large quantities could affect the major sea currents in the Atlantic Ocean. The ocean currents of Atlantic in fact, act as a heat conveyer of the planet regulating the global climate. If the heat conveyer is interrupted, the northern hemisphere would plunge into an ice age and the southern hemisphere will be facing severe drought. In general, global warming is likely to make the weather more unpredictable in the coming years.

3.4.1 Impact on Agriculture and Vegetation

The most obvious effect of climate change will be on agriculture. Because CO_2 is a natural fertilizer, the plants will grow larger and faster with increasing CO_2 in the atmosphere. The abnormal fast growth results in increase of yield but the soil fertility goes down at a very fast rate. Vegetation changes due to climatic change would affect the hydrologic cycle. The biggest impact of CO_2 induced climatic change would be changing precipitation form lead to overall lower rainfall amount or drought during growing season with increased frequency and severity. However, the rise in atmospheric CO_2 should cause increase in photosynthesis, growth and productivity of the earth's vegetation. Thus the change in climate on vegetation has less adverse impact. Higher temperature could increase forest susceptibility to fire, disease and insect damage.

3.4.2 Impact on Society

The social and economic characteristics of a society have also been shaped largely by adapting to the seasonal and year to year patterns of temperature and rainfall. Human society is highly dependent on the earth's climate pattern and human adaptations determine the availability of food, fresh water and other resources for sustaining life

3.4.3 Impact on Water Resources

Due to changes in precipitation pattern and increased evaporation the quality and quantity of water available for drinking, irrigation, industrial use, electric generation, aquatic life, etc., are significantly affected.

3.4.4 Impact on Coastal Resources and Oceans

An estimate of 50 cm rise in sea level by the year 2100, could inundate more than 8000 Km of dry land. Oceans can provide sources for the increased water vapour because of the earth's increased temperature. On the other hand, the thermal holding capacity of the oceans would delay and effectively reduce the observed global warming. In addition, oceans play an important role in the global green house gas budgets. The ocean biota, primarily phytoplankton is believed to remove at least half of the anthropogenic CO_2 added to the atmosphere. The ocean sink of CO_2 is called 'Biological CO_2 Pump'.

3.4.5 Impact on Health

Changing pattern of temperature and precipitation may produce new breeding sites for pests, shifting the range of infectious diseases. Heat stress mortality could increase due to higher temperature over longer periods.

3.4.6 Increase in Clouds and Water Vapour

Global warming will lead to an increase in the amount of water vapour in the atmosphere and because water vapour is a powerful green house gas, lead to an increase into the warming. However, tropical storm clouds reach higher in the atmosphere under warmer conditions. Then the clouds would produce more rain thus adding less water vapour to the middle troposphere.

3.4.7 Control and Prevention of Global Warming

The major steps to be taken for the reduction of green house gases includes, improving the energy efficiency of electric generation, as well as switching to less polluting fossil fuels. Following are some of the suggestions to prevent global warming.

(i) Reduction and elimination of green house gases emission that is disturbing the climate. Use of Clean energy technologies like wind turbine, solar panels and hydrogen fuel cells which are continually improving, becoming more efficient, economical and capable of competing with polluting gas and coal power plants.

(ii) Biofuels including ethanol and bio-diesel could substantially cut down the CO_2 emission from automobiles.

(iii) Sustainable farming and forestry techniques look up carbon in plants and soils and provide new revenues to rural communities.

(iv) Besides protecting the climate, CO_2 emission control techniques dramatically reduce air pollution provide communities with higher quality of life and climate.

(v) Energy Conservation and produce energy that causes no environmental damage with cost less than building new power plants. They lower electricity bills and reduce constraints on energy systems.

(vi) Carbon Sequestration by growing Micro-algae near power plants

3.5 Kyoto Protocol

The Kyoto protocol is a legally binding international agreement to reduce green house gas emissions. It was initially negotiated during a meeting held at Kyoto, Japan in 1977. The protocol commits in industrialized countries to reducing emissions of six green house gases by 5% before 2012.

3.6 Global Dimming

In contrast to global warming there is another phenomenon called 'Global Dimming'. Scientists have observed that 2-4% reduction in the amount of solar radiation reaching the earth's surface, due to increase in cloud cover aerosols and particulates in the atmosphere. Higher temperature leads to an increased cloud cover. The scattered light through the clouds boosts the plant's adsorption of CO_2 and photosynthesis process. Thus global dimming is a process working against global warming to some extent.

3.7 Ozone Layer Depletion

Ozone layer was discovered by French physicists CHARLES FABRY and HENRI BUISSON in 1913.Its properties were explored in detail by G.M.B.DOBSON, a British Meteorologist. Dobson established a world wide network of ozone monitoring stations which operate even today. The total amount of zone in a column overhead is measured in "DOBSON Unit" (DU), 1DU=0.01mm. The average thickness of ozone layer in stratosphere is approximately 300DU.

Although the concentration of ozone is the ozone layer is very small, it is vitally important to life because it absorbs biologically harmful ultra violet (UV) radiation emitted from the Sun. UV radiation is divided into three categories based on its wave length, ie., UV-A, UV-B, UV-C.

(i) Most of the UV-A (315 to 400nm) reaches the surface this radiation is significantly less harmful, although it can potentially cause genetic damage.

(ii) UV-B (280 to 315nm) radiation is the main cause of Sun burn, excessive exposure can also cause genetic damage, resulting in problems such as Skin cancer. It rapidly damages biota of all types.

(iii) UV-C < 280nm, the ozone layer is very effective at screening out UV-B, for radiation with a wave length of 290nm, the intensity at Earth's surface is 350 million times weaker at the top of the atmosphere.

3.7.1 Formation of Ozone in the Atmosphere

Atomic oxygen O, oxygen molecules O_2 and Ozone O_3 are involved in the ozone-oxygen cycle. Ozone is formed in the Stratosphere when oxygen molecules dissociate after absorbing the ultraviolet photon whose wave length is shorter than 240nm.This produces two oxygen atoms. The atomic oxygen then combines with O_2 to create ozone O_3. The process O_3 generation and splitting repeats as per the equations below.

- Ozone absorbs uv radiations and is broken into atomic and molecular oxygen.

$$O_2 \xrightarrow{uv-C} 2O$$

$$O_3 \rightarrow O + O_2$$

- The products formed combine again to form ozone

$$O + O_2 \rightarrow O_3$$

- Ozone molecules absorb UV light between 310 and 200nm, following which ozone splits into a molecule of O_2 and O and hence a dynamic equilibrium is set up due to which the concentration of ozone in the atmosphere remains constant.

$$O_3 + O \rightarrow 2O_2$$

- The ozone layer protects the earth from the harmful uv radiations.
- If the concentration of ozone is reduced (ozone depletion), the concentration of uv radiations reaching the earth increases.
- This leads to irritation of the eyes, skin cancer and damage to immune system in human beings.
- In agriculture it causes decrease in productivity.

3.7.2 Ozone Hole

Ozone layer a region of the atmosphere from 19 to 48 km above the earth's surface. Certain human produced pollutants lead to destroy the stratosphere ozone and causing an imbalance between formation and dissociation of ozone. This decrease in the ozone level is called depletion or thinning of ozone layer or zone hole.

3.7.3 Causes of Ozone Depletion

Ozone can be destroyed by a number of free radical catalyst, like hydroxyl (OH), the nitric oxide (NO), atomic chlorine (Cl) and Bromine (Br). All of these are generated by both natural and anthropogenic (man made) sources.

- Chlorofluorocarbons (CFCs) are used as refrigerants, aerosols and as industrial solvents.

- CFCs are noncombustible and volatile. They reach the atmosphere and are broken down into chlorine free radicals by uv radiations.

$$CF_2Cl_2 \xrightarrow{\text{uv - C}} {}^*CF_2Cl + {}^*Cl$$

The chlorine free radical brings about the degradation of ozone

$$^*Cl + O_3 \rightarrow {}^*ClO + O_2$$

$$^*ClO + O \rightarrow {}^*Cl + O_2$$

- Thus CFCs reduce the concentration of ozone in the
- atmosphere causing ozone hole.

3.7.4 Consequences and Harmful Effects

- Since the ozone layer absorbs UV-B light from the Sun, ozone layer depletion is expected to increase surface UV-B levels, which could lead to damage, including increase in skin cancer.
- Scientists have estimated that a one percent decrease in stratospheric ozone would increase the incidence of skin cancers by 2%.
- A direct correlation has been observed between cataract formation in eyes and UV radiations.
- An increase of UV radiation would also decrease photosynthesis in plants & affect crops yields like rice.
- The temperature on the earth's surface is raised and this leads to global warming.
- Lower trophic level organisms shall be the worst sufferers as they have a simple cell wall for their protection against UV radiation. With the primary tropic levels drastically impaired the entire ecosystems could collapse.

3.7.5 Control of Ozone Depletion

Ozone depletion can be controlled by using hydrochlorofluorocarbons and hydrofluoroalkanes in place of CFCs. These contain more hydrogen in their molecule and undergo oxidation readily.

3.8 Acid Rain

Acid rain is a form of air pollution in which airborne acids produced by electric utility plants and other sources fall to Earth in distant regions. The term was first coined by ROBERT ANGUS SMITH in the year 1852.

The major contributors, called PRECURSORS to the acid are the common air pollutants, like Sulphur dioxide and Nitrogen oxides.Through a variety of chemical reactions the gases form Sulphuric acid and Nitric acid, which are the two acids responsible for the acid rain.

3.8.1 Acid Rain Formation

Nitric oxide can react with oxygen O_2 to form nitrogen dioxide which can be broken down again by Sunlight(h_V)to give Nitric oxide and an oxygen radical (O).

$$2NO + O_2 = 2N_2 \qquad NO_2 + h_V = NO+O$$

The oxygen radical then enables the formation of Ozone (O_3)

$$O + O_2 = O_3$$

The presence of ozone causes the formation of more nitrogen dioxide by its reaction with nitric oxide.

$$NO + O_3 = NO_2 + O_2$$

Or, the oxygen radical reacts with water to give the hydroxyl radical (OH)

$$O + H_2O = 2OH.$$

This radical then reacts with nitric oxide to give nitrous acid (HNO_2) and nitrogen dioxide to give nitric acid (HNO_3). It also combines with Sulphur dioxide to produce Sulphuric acid

$$HO + NH = HNO_2$$

$$NH_2 + HO = HNO_3$$

$$SO_2 + 2HO = H_2SO_4$$

3.8.2 Sources of Acid Rain

While Nitric oxide and Sulphur dioxide are produced biogenically (in nature), there are major anthropogenic (man made) sources of both these polluting gases. Sometimes, natural production of the gases is much higher than human production, but these natural emissions tend to be spread over large area, dispersing their effects, while the man – made emissions are concentrated around the source of their production.

3.8.3 Biogenic Sources (or Natural Sources)

Volcanic eruptions and decay of organic matter produce significant amounts of Sulphur dioxide. Nitrogen oxides are also generated by push fires as well as by microbial process (in Soil) and lightning discharges.

3.8.4 Anthropogenic Sources (or Man made Sources)

Nitrogen oxides are produced mainly from the burning of fossil fuels such as Diesel and petrol in automobiles and from power stations burning coal.

Sulphur dioxide is formed primarily in the burning of (Sulphur containing) Coal, fossil fuels and in metal smelters.

3.8.5 Control of Acid Rain

The most effective way to reduce the incidence of acid deposition is to reduce the emission of its causes – The "PRECURSORS", nitrogen oxides and Sulphur dioxide.

(a) ***Nitrogen oxide reduction:*** The main method of lowering the levels of nitrogen oxides is by a process known as "Catalytic reduction". Catalytic reduction is used in Industry & in motor vehicles.

In motor vehicles the Catalytic converter will convert much of the nitric oxide from the engine gases to the nitrogen and oxygen. Nitrogen is not there in the actual fuels or power stations. It is introduced from the air when combustion occurs. Using less air in combustion can reduce emissions of nitrogen oxides.

Temperature also has an effect on emission. Lower the temperature of combustion, lower will be the production of nitrogen oxides.

Temperatures can be lowered by using processes such as two stage combustion and flue gas recirculation water injection or by modifying the design of the burner.

(b) ***Sulphur dioxide reduction:*** There are several method to lower the Sulphur dioxide emission from Coal – fired stations. Simplest of the lot is using Coal with low Sulphur content and physical coal cleaning. Most Complex is by the process of "FLUE GAS

DESULPHURISATION" and "FLUIDISED BED COMBUSTION".

(c) ***Flue gas desulphurization:*** In this method the Sulphur dioxide (flue gas) is absorbed using lime stone. This method is the most effective of removing Sulphur dioxide. The process generates Solid wastes (Calcium Sulphate, $CaSO_3$ and $CaSO_4$) which require disposal.

$$CaCO_3 \text{ (limestone)} + SO_2 = CaSO_3 + CaSO_4 + CO_2 + H_2O.$$

(d) ***Fluidized bed combustion:*** In this process, coal is crushed and passed into a fluidized "bed" for combustion. The bed consists of fine particles of an absorbent material such as lime stone. Hot air is passed through it and this causes the particles to behave as through they are a fluid. The sulphur dioxide can then be absorbed by the lime stone particles in the bed. Fluidized bed combustion can be operated at lower temperatures and therefore produce less nitrogen oxide, but once again, solid waste is created and requires disposal.

3.8.6 Harmful Effects of Acid Rain

The acids in the acid rain can react chemically with any object they contact. Acids are corrosive chemical that react with other chemical by giving up hydrogen atoms. Acid rain or acid deposition has an adverse effect on environmental eco system as well as humans, animals, buildings, textiles. etc.

(a) ***Soil:*** Acid rain dissolves in Soil and washes away nutrients needed by the plants. It can also dissolve toxic substances such as aluminum & mercury, releasing these toxins to pollute water or to poison plants that absorb them.

Trees: Removal of useful nutrients from the soil, acid rain slows the growth of plants, particularly trees. It also attacks trees more directly by eating holes in the waxy coating of needles & leaves, causing brown dead spots.

Acid rain has been blamed for the decline of Spruce forests on the highest ridges of Apalachian Mountains in the eastern United States. In the black forest of South Western Germany, half of the trees are damaged from the acid rain.

(b) ***Agriculture:*** Most farm crops are less affected by acid rain than the forest. Farmers can prevent acid rain damage by monitoring the condition of the soil and, when necessary, adding crushed lime stone to the soil to neutralize acid.

(c) ***Surface water:*** Acid rain falls into streams, lakes and marshes. Due to this the water life is destroyed. All Norway's major rivers have been damaged by acid rain, severely reducing the fish life.

(d) ***Plants and Animals:*** The effects of acid rain on wild life can be far reaching, if a population of one plant or animal is adversely affected by acid rain, animals that feed on that organism may also suffer ultimately an entire ecosystem may become endangered. Land animals dependent on aquatic organisms are also affected.

(e) ***Man made structure:*** Acid rain and dry deposition of acidic particles damage building, statues, automobiles, and other structures made of stone metal or any other material exposed to weather for long periods. Parthenon in Greece and the Taj- Mahal in India, are deteriorating due to acid deposition.

(f) ***Human health:*** Acidification of Surface water cause little direct harm to human health, it is safe to swim in even the most acidified lakes.

(g) ***In the air:*** acids join with other chemicals to produce urban smog, which can irritate the lungs an make breathing difficult, especially for people with respiratory diseases. Solid particles of sulphates can damage the lungs.

3.8.7 Acid Rain and Global Warming

Acid pollution has one surprising effect that may be beneficial. Sulphates in the upper atmosphere reflect some sunlight out into the space, and thus tend to slow down global warming.

Review Questions

1. Define Pollution. What are the types of pollutants?
2. What are the types of Air pollutants?
3. How will you control air pollution?
4. What are the sources and effects of air pollutants?
5. Name few Green-House gases.
6. Explain the term Global warming. What are the sources of Green House gases in the atmosphere
7. Mention few practices to reduce global warming.
8. Explain acid-rain and its effects.
9. Explain the Mechanism of ozone layer depletion
10. What are the effects of Ozone-Layer depletion?

CHAPTER 4

Sound Pollution

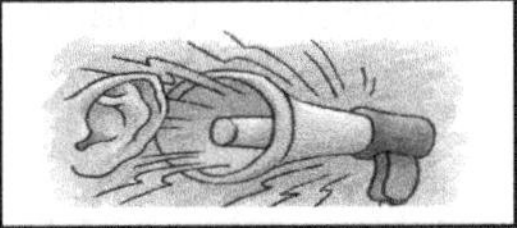

Sound Pollution: Causes, Controlling measures, Measurement of sound pollution (deciblage), Industrial and Non – industrial

4.1 Noise Pollution

Definition: Noise Pollution or Sound Pollution is when the exposure of people or animals to levels of sounds that are annoying, stressful, or damaging to the ears. Although loud and frightening sounds are part of nature, only in recent centuries has much of the world become urban, industrial, and chronically noisy. The machines used in factories make noise throughout the day, and this disturbs the peaceful atmosphere in the vicinity, as machines used without proper covering lead to sound pollution. Noise is an important environmental pollutant like air, water and soil. It destroys bridges and produces cracks in buildings. The noise can cause skin and mental diseases

4.2 Characteristics of Sound: Sound Vs Noise

A vibrating source produces vibration into the medium in which it is placed. These vibrations are propagated as waves in the form of pressure variations and are termed as acoustic waves. If they fall within the range capable of exciting the sense of hearing, they are called as sound waves. An acoustic waves travels in a given medium at a constant velocity. When the level of the sound becomes objectionable, it is called noise. Thus in general, the sound can be referred as a physical or mechanical disturbance capable of being detected by the human ear. The human ear can detect sounds from 20 Hz to 20,000 Hz. The frequencies most important for understanding normal speech like between 300 Hz. to 5,000 Hz. *Noise can be described as sound without agreeable musical quality or as an unwanted or undesired sound. Thus noise can be taken as a group of laud, non harmonious sounds or vibrations that are unpleasant and irritating to ear.*

4.3 Causes & Sources of Sound Pollution

Noise pollution like other pollutants is also a by- product of industrialization, urbanizations and modern civilization. Broadly speaking, the noise pollution has two sources, i.e. ***Industrial and Non-industrial.***

Industrial source includes the noise from various industries and big machines working at a very high speed and high noise intensity.

Non- industrial source of noise includes the noise created by transport/vehicular traffic and the neighborhood

Most noise pollution comes from machines, especially automobiles roads traffic, railroads, trucks, and aircraft. Construction equipment, farm machines, noise in buildings, and consumer products and the din of machinery inside factories can be dangerously loud. Some home appliances, shop tools, lawnmowers, and leaf blowers can also be noisy, as are guns, firecrackers, and some toys. Even music, when played at very high volume, particularly through personal headphones, is as damaging to the ears as a roaring chain saw.

4.3.1 Road Traffic Noise

In the city, the main sources of traffic noise are the motors and exhaust system of autos, smaller trucks, buses, and motorcycles. This type of noise can be augmented by narrow streets and tall buildings, which produce a canyon in which traffic noise reverberates.

4.3.2 Air Craft Noise

Now-a-days , the problem of low flying military aircraft has added a new dimension to community annoyance, as the nation seeks to improve its nap-of the- earth aircraft operations over national parks, wilderness areas, and other areas previously unaffected by aircraft noise has claimed national attention over recent years.

4.3.3 Noise from railroads

The noise from locomotive engines, horns and whistles, and switching and shunting operation in rail yards can impact neighboring communities and railroad workers. For example, rail car retarders can produce a high frequency, high level screech that can reach peak levels of 120 dB at a distance of 100 feet, which translates to levels as high as 138, or 140 dB at the railroad worker's ear.

4.3.4 Construction Noise

The noise from the construction of highways, city streets, and buildings is a major contributor to the urban scene. Construction noise sources include pneumatic hammers, air compressors, bulldozers, loaders, dump trucks (and their back-up signals), and pavement breakers.

4.3.5 Noise in Industry

Although industrial noise is one of the less prevalent community noise problems, neighbors of noisy manufacturing plants can be disturbed by sources such as fans, motors, and compressors mounted on the outside of buildings Interior noise can also be transmitted to the community through open windows and doors, and even through building walls. These interior noise sources have significant impacts on industrial workers, among whom noise- induced hearing loss is unfortunately common.

4.3.6 Noise in building

Apartment dwellers are often annoyed by noise in their homes, especially when the building is not well designed and constructed. In this case, internal building noise from plumbing, boilers, generators, air conditioners, and fans, can be audible and annoying. Improperly insulated walls and ceilings can reveal the sound of-amplified music, voices, footfalls and noisy activities from neighboring units. External noise from emergency vehicles, traffic, refuse collection, and other city noises can be a problem for urban residents, especially when windows are open or insufficiently glazed.

4.3.7 Noise from Consumer Products

Certain household equipment, such as vacuum cleaners and some kitchen appliances have been and continue to be noisemakers, although their contribution to the daily noise dose is usually not very large.

4.4 Harmful Effects of Noise

4.4.1 Effect on Human Being

Noise has always been with the human civilization but it was never so obvious, so intense, so varied & as pervasive as it is seen in the last of this century. Noise pollution makes men more irritable. The effect of noise pollution is multifaceted & inter related. The effects of Noise Pollution on Human Being, Animal and property are as follows:

(i) *It decreases the efficiency of a man:* Regarding the impact of noise on human efficiency there are number of experiments which print out the fact that human efficiency increases with noise reduction. A study by Sinha & Sinha in India suggested that reducing industrial booths could improve the quality of their work. Thus human efficiency is related with noise.

(ii) *Lack of concentration:* For better quality of work there should be concentration, Noise causes lack of concentration. In big cities, mostly all the offices are on main road. The noise of traffic or the loud speakers of different types of horns divert the attention of the people working in offices.

(iii) *Fatigue:* Because of Noise Pollution, people cannot concentrate on their work. Thus they have to give their more time for completing the work and they feel tiring

(iv) *Abortion is caused:* There should be cool and calm atmosphere during the pregnancy. Unpleasant sounds make a lady of irritative nature. Sudden Noise causes abortion in females.

(v) *It causes Blood Pressur*e: Noise Pollution causes certain diseases in human. It attacks on the person's peace of mind. The noises are recognized as major contributing factors in accelerating the already existing tensions of modern living. These tensions result in certain disease like blood pressure or mental illness etc.

(vi) *Temporary of permanent Deafness*: The effect of nose on audition is well recognized. Mechanics, locomotive drivers, telephone operators etc. All have their hearing. Impairment as a result of noise at the place of work. Physicist, physicians & psychologists are of the view that continued exposure to noise level above. 80 to 100 db is unsafe, loud noise causes temporary or permanent deafness.

4.4.2 Effect on vegetation

Poor quality of Crops:- Now is well known to all that plants are similar to human being. They are also as sensitive as man. There should be cool & peaceful environment for their better growth. Noise pollution causes poor quality of crops in a pleasant atmosphere.

4.4.3 Effect on animal

Noise pollution damages the nervous system of animal. Animal looses the control of its mind. They become dangerous.

4.4.4 Effect on property

Loud noise is very dangerous to buildings, bridges and monuments. It creates waves which struck the walls and put the building in danger condition. It weakens the edifice of buildings.

4.5 Controlling Measures of Noise Pollution

A. *Industrial Noise*: Control of noise (i.e. its prevention and reduction) is a system related problem. This system is composed of the following components

- The source of noise
- The path of sound propagation; and
- The receiver of noise.
    - The source is part of the system that produces the acoustic energy. The source should be considered as a group of noise generators which present different physical characteristics randomly distributed in space and time.
    - The acoustic energy produced by the source is transmitted to the environment in which it is propagated, and which may be a solid structure or air.

- The third component of the system (i.e the receiver of sound) may be a worker operating his machines.

Noise prevention and reduction measures should be aimed at the following goals:

- Controlling source of noise
- Precluding the propagation, amplification and reverberation of noise
- Isolating the workers.

Evidently, reduction of noise at the source is the most rational method of noise control. If all machines and all prime movers were sufficiently silent, there would not be many problems of excessive noise left for solution.

B. Community Noise:

(i) The traffic volume should be reduced by diversion of traffic, and use of horn should be banned.

(ii) Residences should not be allowed to grow nearby industrial areas.

(iii) There should be plenty of trees and bushes in open spaces, houses and lanes.

(iv) At the time of festive occasions, as practiced in some western countries, community fireworks could be displayed in lieu of individual celebrations, which will be a good way of enjoying the spirit of the occasion while keeping it safe and pollution free, thus lowering the noise, which mars the celebrations.

(v) Also, as a measure of awareness drive, slums, schools, and residential colonies covering all zones should be visited to address both youngsters and elders on the need of adhering to the relevant decibels and smoke emission levels in a bid to make festivals noise and pollution free,

(vi) Proper monitoring noise levels and ambient air quality.

C. Legal Control:

(a) *Constitution of India*

Right to Life: Article 21 of the Constitution guarantees life and personal liberty to all persons. It is well settled by repeated pronouncements of the Supreme Court that right to life enshrined in Article 21 is not of mere survival or existence. It guarantees a right of persons to life with human dignity. Any one who wishes to live in peace, comfort and quiet within his house has a right to prevent the noise as pollutant reaching him.

Right to Information: Everyone has the right to information know about the norms and conditions on which Govt. permit the industry which effect the environment.

Right to Religion and Noise: Right to religion does not include right to perform religious activities on loud speaker and electronic goods which produce high velocity of noise.

Directive Principal of State Policy: The state has the object to make the environment pollution free.

Fundamental Duties: every citizen of the country has the fundamental duty to clean the environment.

(b) *Cr.P.C. Section 133*

Here Section 133 is of great importance. Under Crpc. Section 133 the magisterial court have been empowered to issue order to remove or abate nuisance caused by noise pollution Sec 133 empower an executive magistrate to interfere and remove a public nuisance in the first instance with a conditional order and then with a permanent one. The provision can be utilized in case of nuisance of environment nature. He can adopt immediate measure to prevent danger or injury of a serious land to the public. For prevention of danger to human life, health or safety the magistrate can direct a person to abstain from certain acts.

(c) *I.P.C. Public Nuisance 268-295*

Chapter IV of Indian Penal code deals with offences relating to public health, safety, decency , morals under Sections 268, 269, 270, 279, 280, 287, 288, 290 291 294. Noise pollution can be penalized with the help of above section. Private remedies suits in the area may related to public nuisance under A299. This article punishment in case of Public nuisance law of torts covers. A person is guilty of public nuisance who does any act or is guilty of an illegal omission which causes any common injury, danger, or annoyance to the pubic or to the people in general who dwell or occupy property in the vicinity or which must necessarily cause injury, obstruction danger or annoyance to persons who may have occasion to use any public right. A common nuisance is not excused on the ground that it causes some convenience or advantage. Who ever commits a public nuisance in any case not otherwise punishable by this code, shall be punished with fine, which may extend to Rs. 200.

(d) *Law of Torts Noise pollution is considered as civil wrong*

Under law of torts, a civil suit can be filed claiming damages for the nuisance. For filing a suit under law of torts a plaintiff is required to comply with some of the requirement of tort of nuisance which are as follows:-

(i) There should be reasonable interference.

(ii) Interference should be with the use & enjoyment of land.

(iii) In an action for nuisance actual damage is required to be proved. As a general rule either the presence or absence of malice does not matter. But in some cases deviation from the rule has been made.

In Christe Vs Davey The extent of noise & the amount of disturbance caused there by was ignored & it was held that the noise which arose due to the practice of lawful profession, & without any malice, could not be considered to be actionable nuisance.

In Hollywood Silver Fox Farm Ltd. Vs Emmett It was held that presence of malice was a factor in determining liability for noise amounting to nuisance. The court said that even on his won land was nuisance, & the defendant was liable in damages.

(e) *Factories Act Reduction of Noise and Oil of Machinery*

The Factories Act does not contain any specific provision for noise control. However, under the Third Schedule Sections 89 and 90 of the Act, noise induced hearing loss, is mentioned as notifiable disease. Similarly, under the Modal Rules, limits for noise exposure for work zone area have been prescribed.

(f) Motor Vehicle Act. Provision Relation to use of horn and change of Engine

In Motor vehicle Act rules regarding use horns and any modification in engine are made.

(g) Noise Pollution Control Rule 2000 under Environment Protection Act 1996

Further for better regulation for noise pollution There are The Noise Pollution (Regulation and Control) Rules, 2000 – in order to curb the growing problem of noise pollution the government of

India has enacted the noise pollution rules 2000 that includes the following main provisions:-

- The state government may categories the areas in the industrial or commercial or residential.
- The ambient air quality standards in respect of noise for different areas have been specified.
- State government shall take measure for abatement of noise including noise emanating from vehicular movement and ensure that the existing noise levels do not exceed the ambient air quality standards specified under these rules.
- An area not less than 100 m around hospitals educations institutions and court may be declare as silence are for the purpose of these rules.
- A loud speaker or a public address system shall not be used except after obtaining written permission from the authority and the same shall not be used at night. Between 10 pm to 6 am.
- A person found violating the provisions as to the maximum noise permissible in any particular area shall be liable to be punished for it as per the provision of these rules and any other law in force.

4.6 Noise Pollution Standards

The Noise Pollution (Regulations & Control) Rules 2000 prescribe the ambient air quality standards in respect to noise in industrial, commercial, residential and silence zone areas as below-

S. No.	Area	Noise Level (Leq in dBA)	
		Day	Night
1.	Industrial	75	70
2.	Commercial	65	55
3.	Residential	55	45
4.	Silence Zone	50	40

A "decibel" is a unit in which noise is measured. "A", in dB (A) Leq: It is energy mean of the noise level, over a specified period. Or it denotes the time weighted average of the level of sound in decibels on scale which is relatable to human hearing.

4.7 Measurement of Sound Pollution

Broadly speaking, Intensity is the energy of sound waves and Frequency is the number of times per second the sound waves hit the ear of man On the basis of the above definition, the loudness which is more commonly understood source of noise, is in fact Combination of Intensity and Frequency can be measured in DECIBELS. Noise intensity is measured in decibel units. The decibel scale is logarithmic; each 10-decibel increase represents a tenfold increase in noise intensity. It is to be remembered that the threshold at the normal hearing is 20-25 decibels and of normal conversation is 60 decibels. It has also been noticed that speech interference occurs at 75 decibels and definite annoyance begins at 80 decibels. The motor- activities disturbed at 90 decibels and physiological disturbance occurs beyond 120 decibels definite pain occurs at 140 decibels.

Intensity – The level of sound is usually expressed in terms of the Sound Pressure Level (SPL) in decibels, which is defined as

$$SPL = 20 \log \ 10 \ P/PO \ dB$$

where P is the pressure variation mea sured in N/ m^2 and PO is the standard reference pressure taken as 2×10^{-5} N/m2

Frequency – Frequency of a sound wave is the number of times it repeats itself in each sound (i.e the rapidity, with which the pressure fluctuations occur)

Human beings generally have the ability to hear sounds in the frequency range 20 to 20,000 Hz (1 Hz = 1 cycle per second) The point at which the limit of the hearing threshold i.e., O dB, the sound pressure is equal to the standard reference pressure of 2×10^{-5} N /m^2.

Human hearing is most sensitive to frequencies in the range 500 to 6000 Hz and less sensitive both at lower and higher frequencies.

Review Questions

1. Briefly describe the sources of noise pollution
2. Write short notes on Noise Pollution Standards
3. Define Sound Pollution
4. Explain the Harmful effects of Noise Pollution.
5. Briefly explain the methods to control Noise Pollution

CHAPTER 5

Water Pollution

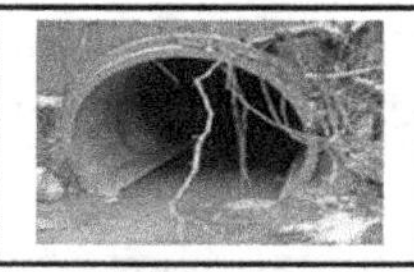

***Water Pollution*:** Pollutants in water, Adverse effects. Treatment of Domestic and Industrial water effluent.

5.1 Water Pollution

Introduction: Water exists in three states: solid, liquid and gaseous. The important sources of water are (i) rain water, (ii) ground water and (iii) sea water. Rain water carries the washed out minerals, salts and organic matter from the earth's surface and stores them in ponds, lakes and rivers. It seeps into underground and is stored as ground water. Sea water is highly alkaline due to the presence of dissolved salts. The natural water contains numerous organisms and dissolved gases (ex: oxygen), which is essential for aquatic organisms. The pure water is one which is free from organisms. Next to air, water is the most important substance for the existence of life on the earth. Today water resources have been the most exploited natural systems since man's existence on the earth. Pollution of water bodies is increasing steadily due to rapid population growth, industrial proliferations, urbanizations, increasing living standards and wide spread human activities Groundwater, river, seas, lakes, ponds and streams are founding it more and more difficult to escape from pollution.

Water is required mainly for drinking and cooking, also for industry, agriculture and many other activities.

Pollution of water implies that it contains a lot of inorganic and organic substances introduced by human activities, which change its quality, not suitable for any purposes and also harmful for living organisms.

(OR)

Any alteration in physical, chemical or biological properties of water, as well as the addition of any foreign substance makes it unfit for health and which decreases the utility of water, is known as water pollution

- The substances which cause pollution are called pollutants and the common pollutants which are present in water are (i) Suspended solids (ii) Organic matter, (iii) Inorganic pollutants, (iv) Oil, etc.
- Turbidity in water is mainly due to; (i) finely divided undissolved solids, clay, slit; (ii) colloidal particles and (iii) organic matters. Turbidity gives unsightly appearance. When it is used in industries, it causes problems in functioning of equipments, boilers, etc. This can be removed from water by applying proper treatments like settling, coagulation (by using alum) and filtration.
- Organic pollutants include domestic and animal sewage, biodegradable organic compounds, industrial wastes, synthetic pesticides, fungicides, herbicides, detergents, oil, grease, pathogenic microorganisms, etc. It results in rapid depletion of dissolved oxygen of water and thus such water becomes harmful for aquatic lives. Organic matter present in water can be removed by using chlorination, coagulation and ultra filtration processes.
- Inorganic pollutants consist of mineral acids, inorganic salts, finely divided metals, cyanides, sulphates, nitrates, organometallic compounds, etc.
- Oil and grease constitutes important water pollutants. These substances coat ion exchange resin, causes premature exhaustion of beds. It can be removed by coagulation with alum.

5.2 Main Sources of Water Pollution

Water pollutants result from many human activities. Pollutants from industrial sources may pour out from the outfall pipes of factories or may leak from pipelines and underground storage tanks. Polluted water may flow from mines where the water has leached through mineral-rich rocks or has been contaminated by the chemicals used in processing the ores. Cities and other residential communities contribute mostly sewage, with traces of household chemicals mixed in. Sometimes industries discharge pollutants into city sewers, increasing the variety of pollutants in municipal areas. Pollutants from such agricultural sources as farms, pastures, feedlots, and ranches contribute animal wastes, agricultural chemicals, and sediment from erosion. Water pollution can also be caused by other types of pollution. For example, sulfur dioxide from a power plant's chimney begins as air pollution. The polluted air mixes with atmospheric moisture to produce airborne sulfuric acid, which falls to the earth as acid rain. In turn, the acid rain can be carried into a stream or lake, becoming a form of water pollution that can harm or even eliminate wildlife. Similarly, the garbage in a landfill can create water pollution if rainwater percolating through the garbage absorbs toxins before it sinks into the soil and contaminates the underlying groundwater (water that is naturally stored underground in beds of gravel and sand, called aquifers).

Sources of surface water pollution are generally grouped into two categories based on their origin:

- Point source pollution: Point source pollution refers to contaminants that enter a waterway through a discrete conveyance, such as a pipe or ditch. Examples of sources in this category include discharges from a sewage treatment plant, a factory, municipal sewer systems, as well as water from construction sites.
- Non-point source pollution: Non-point source (NPS) pollution refers to diffuse contamination that does not originate from a single discrete source. NPS pollution is often accumulative effect of small amounts of contaminants gathered from a large area. The leaching out of nitrogen compounds from agricultural land which has been fertilized is a typical example.
- Major Sources of Water pollution are (i) domestic and municipal sewage; (ii) industrial waste; (iii) agricultural waste; (iv) radioactive materials, etc.
- Domestic sewage consists of human excreta, street wastes, organic substances that provide nutrition for bacteria and fungi. It is grey green or grey yellow in color and darkens with time due to decomposition, when becomes stale it develops offensive odor due to evolution of gases like NH_3, H_2S, etc. It is normally turbid due to the presence of suspended solids. Its temperature is slightly higher than ordinary water. These pollutants cause many hazardous effects

on health. Discharge of sewage in river and lakes spreads water borne diseases.

- A pollutant present in industrial waste water damages biological activities and kills many useful organisms. Most of the industrial wastes dissolved in water are particulate in nature and are present at the bottom of the water system. These acts as poison for the aquatic organisms. Further, toxic metals present in industrial effluents are extremely hazardous for living beings.
- Agricultural discharge consists of pesticides, fertilizers, insecticides, etc. In agriculture in order to increase the production and to escape the crops from various diseases, the fertilizers and insecticides are used. Any substance or a mixture of substances which prevents, repels, destroys any pest is called a pesticide. These pollutants contaminate the water and when this is used by human being, affect the oxygen carrying capacity of hemoglobin and consequently causes suffocation and irritation to respiratory and vascular system.
- Radioactive wastes are mainly from atomic explosion and processing of radioactive materials near the source of water. The other sources are waste from hospitals, research laboratories, etc. The radioactive pollutants in water cause serious skin cancer, carcinoma, leukemia, DNA breakage, etc.
- Water pollution by heavy metals: About 70 metallic elements are called heavy metals, as they have atomic numbers of 22 to 92 and atomic weight higher than that of sodium and with a specific gravity of more than 5.0. Only a few of these heavy metals are considered potentially damaging to living systems.

5.3 Types of Water Pollutants

The major water pollutants are chemical, biological, or physical materials that degrade water quality. Pollutants can be classed into five categories, each of which presents its own set of hazards.

5.3.1 Petroleum Products

Oil and chemicals derived from oil are used for fuel, lubrication, plastics manufacturing, and many other purposes. These petroleum products get into water mainly by means of accidental spills from ships, tanker trucks, pipelines, and leaky underground storage tanks. Many petroleum products are poisonous if ingested by animals, and spilled oil damages the feathers of birds or the fur of animals, often causing death. In addition, spilled oil may be contaminated with other harmful substances, such as polychlorinated biphenyls (PCBs).

5.3.2 Pesticides and Herbicides

Chemicals used to kill unwanted animals and plants, for instance on farms, may be collected by rainwater runoff and carried into streams. Some of these chemicals are biodegradable and quickly decay, while others are non-biodegradable and remain dangerous for a long time. When animals consume plants that have been treated with certain non-biodegradable chemicals, they are absorbed into organs of the animals. Many drinking water supplies are contaminated with pesticides from widespread agricultural use. Environmental Protection Agency (EPA), Pune, estimates that more than 14 million Indians drink water contaminated with pesticides.

5.3.3 Hazardous Wastes

Hazardous wastes are chemical wastes that are either *toxic* (poisonous), *reactive* (capable of producing explosive or toxic gases), *corrosive* (capable of corroding steel), or *ignitable* (flammable). If improperly treated or stored, hazardous wastes can pollute water supplies. In 1969 the Cuyahoga River in Cleveland, Ohio, was so polluted with hazardous wastes that it caught fire and burned. In the United States in 1993, about 250 million metric tons of hazardous wastes were produced, and hazardous waste went to 2584 treatment, storage, or disposal sites.

5.3.4 Excess Organic Matter

Fertilizers and other nutrients used to promote plant growth on farms and in gardens may find their way into water. At first, these nutrients encourage the growth of plants and algae in water. However, when the plant matter and algae die and settle underwater, microorganisms decompose them. In the process of decomposition, these microorganisms consume oxygen that is dissolved in the water. Oxygen levels in the water may drop to such dangerously low levels that oxygen-dependent animals in the water, such as fish, die.

5.3.5 Sediments

Sediments or soil particles carried to a streambed, lake, or ocean, can also be a pollutant if it is present in large enough amounts. Soil erosion produced by the removal of soil-trapping trees near waterways, or carried by rainwater and floodwater from croplands, strip mines, and roads, can damage a stream or lake by introducing too much nutrient matter. Sedimentation can also cover streambed gravel in which many fish, such as salmon and trout, lay their eggs.

5.4 Harmful & Health Effects of Water Pollution

5.4.1 Water Borne Diseases

About 1.1 billion people in the world still lack access to safe water for drinking and 2.4 billion people have no basic sanitation. The large majority of people is seriously affected by or dies from preventable water and sanitation related diseases are rural dwelling and the urban poor in the developing countries. Current international estimate of deaths are due to water related diseases which range from 2.2 million to 5 million annually.

(i) ***Classification of Water Related Diseases*:** Water related diseases can be grouped into four general classes: water borne, water-washed, water-based and water related insect vectors. The first three classes are closely linked to people's lack of access to safe water supply.

Disease Classification	Description
Water Borne Diseases	Caused by the consumption of water contaminated by human or animal excreta (feces, urine) containing disease causing organisms such as bacteria, viruses, worms and amoebas.
Water-Washed Diseases (Water Scared Diseases)	Caused by poor personal hygiene and skin or eye contacts with contaminated water and / or insufficient quantities of water for personal hygiene and washing **Ex**: Scabies, trachoma (eye infections), flea, lice, typhus
Water-based Diseases	Caused by parasite found in intermediate organisms living in contaminated water. These diseases are passed on to humans when they drink / wash with it. **Ex**: Dracunculiasis, Schistosomiasis, other helminthes
Water Related Insect-Vector Diseases	Caused by insects, especially flies and mosquitoes that breed in or feed near contaminated water sources. **Ex**: Malaria, dengue, blindness, sleeping sickness, yellow fever.

(ii) Description of Selected Water Borne Diseases

Disease	Description
Diarrhea	It is the most common type of water related illness and is caused by drinking water contaminated with disease causing bacteria, viruses and / or tiny parasites like worms / amoebas from human excreta. People who are sick with this have to defecate more often than usual which results in problem of dehydrations and malnutrition.
Dysentery	It is a more serious form of diarrhea and occurs when contaminated water is used for eating / drinking. The persons' feces will frequently contain blood or mucus. It spreads from person to person.
Cholera	It is a highly contagious diarrhea caused by drinking / eating food of water contaminated with the feces or vomit of an infected person. It can also be spread by dirty hands / flies. Cholera outbreaks commonly occur in crowded slums and in the aftermaths of major diseases where water and sanitation facilities are non-existent / damaged / destroyed. In severe cases rapid loss of body fluids leads to dehydration and shock.
Typhoid Fever	Typhoid is a gut infective caused by food / water contaminated with bacteria found in human excreta, and often occurs in epidemics. This disease results in high fever accompanied with diarrhea or vomiting.
Trachoma (Eye Infection)	It is a chronic form of conjunctivitis (pink eye) that get progressively worse and may last for months or many years. The disease is spread by touch of flies. Trachoma is a major cause of blindness in developing countries.
Malaria	Malaria is a disease caused by the micro organisms that are passed onto people who are bitten by malaria infected mosquitoes. People who suffer from malaria suffer from recurring attacks that cause shivers, fevers and aches.
Schistosomiasis	This disease is caused by blood flukes-tiny worms that begin their lives inside fresh water snails. After being released with water as free-swimming worms they penetrate the skin who are swimming, bathing or washing in contaminated water. Once in the blood stream, the worm cause victims to suffer from fever, pain in the lower abdomen over time this results in liver damage.
Trypanosomiasis (Sleeping Sickness)	This disease is a dangerous infection spread by infected flies commonly found much in woodland especially around water holes where flies breed. People suffer from fever, ache, fatigue and progressive confusion and difficulty in walking and talking.
Dracunchliasis (Guinea Worm)	This disease is resulted by small worms that enter people in bodies when they drink contaminated water. These worms in body can grow upto 50 cm in length or more just under the skin. The adult worm will form a blister on the skin, normally in the lower parts of the body or legs. When the blister pops, the worm will start to come out of the victim's body.

(iii) **Other effects of water-bone diseases:** In addition to the immediate and often devasting health effects of water related diseases, affected individuals cannot work. Meager savings are exhausted, people become poor, cannot be productive in turn results in poverty. Water borne infections hamper absorption of food even when intake is sufficient causing malnutrition.

(iv) **Prevention and solutions:**

(a) ***Water-Borne Diseases:*** Provides wholesome water and good sanitation. Constructing sanitary latrines and treating wastewater to allow for biodegradation of human waste will help to curb diseases caused by pollution.

(b) ***Water-Washed Diseases****:* They can be controlled effectively with better hygiene for which adequate freshwater is necessary.

(c) ***Water-Based Diseases****:* Individuals can prevent infection from water-based diseases by washing vegetables in clean water and thoroughly cooking the food. Practicing filtration with nylon gauge clothes to remove guinea worms. Good hygiene, suitable disposal of human waste.

(d) ***Water Related Insect-Vector Diseases****:* The solution to water related vector diseases would appear to be clear to eliminate the insects that transmit diseases. Putting pesticides, there also have some negative effects. Alternate techniques include using bed nets / introducing predators and sterile insects. Another way is using biological methods and habitat management to reduce / eliminate the natural breeding grounds of the disease vectors. What is important is to have wholesome drinking water to reduce the incidence of diseases and also to reduce malnutrition. Sustainability needs to be addressed by moving away wherever possible from groundwater to surface water resources or groundwater recharge.

(v) Sources and ill effects of heavy metals:

Heavy Metal	Sources	Their effects
Cd	Discharges from electroplating industries, Battery manufacturing units, metallurgical industries, etc.	Gets adsorbed on suspended matter in the water, when it is consumed causes liver and kidney necrosis, increased salivation nausea, acute gastritis, etc.
Hg	Effluents from chloro-alkali industries, pesticide industries, Chemical industries, etc.	Mercury poisoning causes kidney damage, and exhibits the symptoms like numbness in the limbs, muscles, blurred vision leading to blindness, emotional disturbances etc. It also damages brain and nervous system, and paralysis followed by death.
Pb	Electric storage battery industries, petroleum industries, ceramic industries, electric cable insulation, paint industries, plastic industries, pesticides, pipe-manufacturing units, etc.	A cumulative poison causing loss of apatite, constipation, abdominal pain, mental retardation, nervous disorder and brain damage.
CN^-	Metal finishing and cleaning, electroplating, coke ovens and many other industrial processes generate cyanide and discharge as effluent to water bodies.	Cyanide is extremely toxic. Exposure even to small quantities over longer periods causes loss of apatite, dizziness, etc.
NH_3	Ammonia is generated by the biological decay, reduction of nitrates under anaerobic conditions.	In high concentration, it is toxic to fish and other aquatic organisms. It imparts characteristic odor to water.
H_2S	By bacterial reduction of sulphate and decomposition of organic matter.	Causes corrosion, imparts bad odor.

5.5 Domestic Sewage Treatment

Domestic sewage is 99.9% pure water; the other .1% are pollutants. While found in low concentrations, these pollutants pose risk on a large scale. Municipal treatment plants are designed to control conventional pollutants: mainly suspended solids. Well-designed operated systems (i.e., secondary treatment or better) can remove 90 percent or more of these pollutants. Some plants have additional sub-systems to treat nutrients and pathogens. However, most municipal plants are not designed to treat toxic pollutants found in industrial wastewater.The polluted water is characterized by its oxygen demand and solid content. The biological oxygen demand (BOD) measures the level of organic pollution in the sewage water. The sewage must be treated before being discharged into the water bodies. The treatment is carried out in three stages- primary, secondary and tertiary. In waste-water treatment plants, the unit operations and processes described in the previous section are grouped together in a variety of configurations to produce different levels of treatment, commonly referred to as preliminary, primary, secondary and tertiary or advanced treatment (see figure 5.1).

5.5.1 Preliminary treatment

Preliminary treatment prepares waste-water influent for further treatment by reducing or eliminating non-favourable waste-water characteristics that might otherwise impede operation or excessively increase maintenance of downstream processes and equipment. These characteristics include large solids and rags, abrasive grit, odours, and, in certain cases, unacceptably high peak hydraulic or organic loadings.

Preliminary treatment processes consist of physical unit operations, namely screening and comminution for the removal of debris and rags, grit removal for the elimination of coarse suspended matter, and flotation for the removal of oil and grease.Other preliminary treatment operations include flow equalization, septage handling, and odour control methods.

5.5.2 Primary treatment

Primary treatment involves the partial removal of suspended solids and organic matter from the waste-water by means of physical operations such as screening and sedimentation. Preaeration or mechanical flocculation with chemical additions can be used to enhance primary treatment. Primary treatment acts as a precursor for secondary treatment. It is aimed mainly at producing a liquid effluent suitable for downstream biological treatment and separating out solids as a sludge that can be conveniently and economically treated before ultimate disposal. The effluent from primary treatment contains a good deal of organic matter and is characterized by a relatively high BOD.

5.5.3 Secondary treatment

The purpose of secondary treatment is the removal of soluble and colloidal organics and suspended solids that have escaped the primary treatment. This is typically done through biological processes, namely treatment by activated sludge, fixed-film reactors, or lagoon systems and sedimentation.

5.5.4 Tertiary/advanced waste-water treatment

Tertiary treatment goes beyond the level of conventional secondary treatment to remove significant amounts of nitrogen, phosphorus, heavy metals, biodegradable organics, bacteria and viruses. In addition to biological nutrient removal processes, unit operations frequently used for this purpose include chemical coagulation, flocculation and sedimentation, followed by filtration and activated carbon. Less frequently used processes include ion exchange and reverse osmosis for specific ion removal or for dissolved solids reduction.

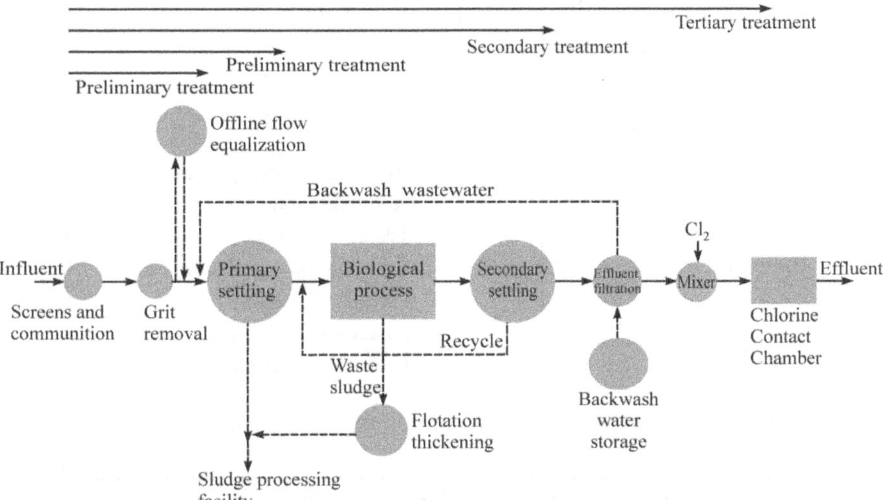

Fig. 5.1 Various treatment levels in a waste-water treatment plant flow diagram

The treated water is of high clarity, free from odor and low BOD, therefore it is nearly equivalent to drinking water.

Waste-water treatment methods are broadly classifiable into physical, chemical and biological processes. Figure 5.2 lists the unit operations included within each category Physical, chemical and biological methods are used to remove contaminants from waste-water. In order to achieve different levels of contaminant removal, individual waste-water treatment procedures are combined into a variety of systems, classified as primary, secondary, and tertiary waste-water treatment.

More rigorous treatment of waste-water includes the removal of specific contaminants as well as the removal and control of nutrients. Natural systems are also used for the treatment of waste-water in land-based applications. Sludge resulting from waste-water treatment operations is treated by various methods in order to reduce its water and organic content and make it suitable for final disposal and reuse.

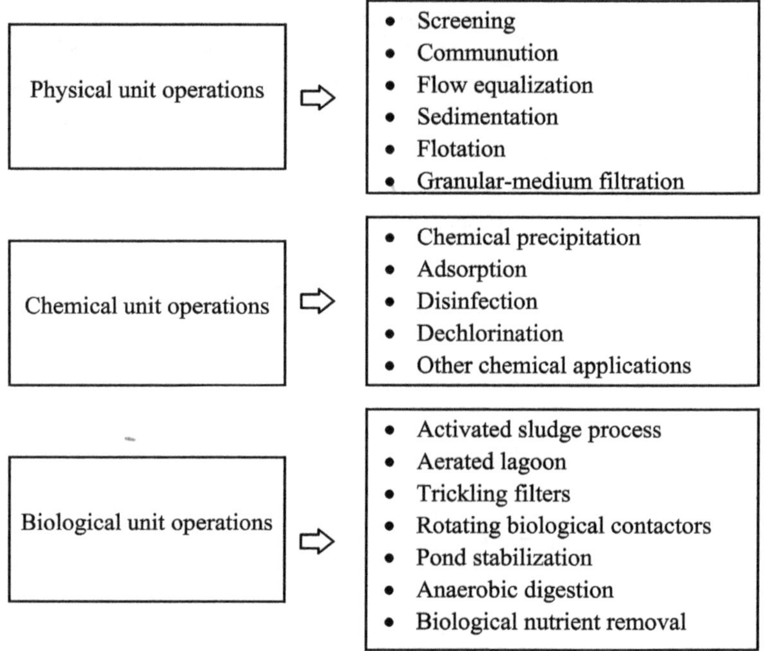

Fig. 5.2 Waste-water treatment unit operations and processes

5.6 Industrial Wastewater Treatment

Most industries generate ordinary domestic sewage that can be treated by municipal treatment plants. However, industries, that generate wastewater with high concentrations of conventional pollutants (such as oil and grease), toxic pollutants (such as heavy metals, volatile organic compounds) or other non-conventional pollutants (such as ammonia), need specialized treatment systems.

Industrial Water Treatment can be classified into the following categories:

1. Boiler water treatment
2. Cooling water treatment
3. Wastewater treatment

Water treatment is used to optimize most water-based industrial processes, such as: heating, cooling, processing, cleaning, and rinsing, so that operating costs and risks are reduced. Poor water treatment lets water interact with the surfaces of pipes and vessels which contain it. Steam boilers can scale up or corrode, and these deposits will mean more fuel is needed to heat the same amount of water. Cooling towers can also scale up and corrode, but left untreated, the warm, dirty water they can contain will encourage bacteria to grow, and Legionnaires' Disease can be the fatal consequence. Also, water treatment is used to improve the quality of water contacting the manufactured product e.g. semiconductors, and/or can be part of the product e.g. beverages, pharmaceuticals, etc. In these instances, poor water treatment can cause defective products. Domestic water can become unsafe to drink if proper hygiene measures are neglected.

In many cases, effluent water from one process might be perfectly suitable for reuse in another process somewhere else on site. With the proper treatment, a significant proportion of industrial on-site wastewater might be reusable. This can save money in three ways: lower charges for lower water consumption, lower charges for the smaller volume of effluent water discharged and lower energy costs due to the recovery of heat in recycled wastewater.

5.6.1 Objectives of industrial water treatment

Industrial water treatment seeks to manage four main problem areas: scaling, corrosion, microbiological activity and disposal of residual wastewater. Boilers do not have many problems with microbes as the high temperatures prevent their growth.

Scaling occurs when the chemistry and temperature conditions are such that the dissolved mineral salts in the water are caused to precipitate and form solid deposits. These can be mobile, like a fine silt, or can build up in layers on the metal surfaces of the systems. Scale is a problem because it insulates and heat exchange becomes less efficient as the scale thickens, which wastes energy. Scale also narrows pipe widths and therefore increases the energy used in pumping the water through the pipes.

Corrosion occurs when the parent metal oxidises (as iron rusts, for example) and gradually the integrity of the plant equipment is compromised. The corrosion products can cause similar problems to scale, but corrosion can also lead to leaks, which in a pressurised system can lead to catastrophic failures.

Microbes can thrive in untreated cooling water, which is warm and sometimes full of organic nutrients, as wet cooling towers are very efficient air scrubbers. Dust, flies, grass, fungal spores and so on collect in the water and creates a sort of "microbial soup" if not treated with biocides. Most outbreaks of the deadly Legionnaires' Disease have been traced to unmanaged cooling towers, and the UK has had stringent Health & Safety guidelines concerning cooling tower operations for many years as have had governmental agencies in other countries.

5.6.2 Disposal of residual industrial waste-waters

Disposal of residual wastewaters from an industrial plant is a difficult and costly problem. Most petroleum refineries, chemical and petrochemical plants have onsite facilities to treat their waste-waters so that the pollutant concentrations in the treated waste-water comply with the local and/or national regulations regarding disposal of waste-waters into community treatment plants or into rivers, lakes or oceans.

Water generated by power plants or manufacturing plants is generally heated. It may be controlled with the help of:

- Cooling ponds, man-made bodies of water designed for cooling by evaporation, convection, and radiation
- Cooling towers, which transfer waste heat to the atmosphere through evaporation and/or heat transfer
- Cogeneration, a process where waste heat is recycled for domestic and/or industrial heating purposes.

5.7 Water Quality Parameters

5.7.1 Biological oxygen demand (BOD)

BOD test is a test for judging water quality. It is done to determine the approximate quantity of oxygen required to biologically stabilize the organic matter present and to determine the size of waste treatment facilities.

It is defined as the amount of oxygen required for the biological oxidation of the organic matter under aerobic conditions at 20°C and for a period of 5 days.

Characteristics of BOD

- It is expressed in parts per million (ppm) or mg/dm^3.

- Larger the concentration of decomposable organic matter, greater is the BOD and consequently more is the nuisance value.
- Strictly aerobic conditions are required.
- Determination is slow and time consuming.

Determination BOD

- The method is based on the determination of dissolved oxygen before and after 5 days period, at 20°C.
- A known volume of sample of sewage is diluted with known volume of water containing nutrients for bacterial growth, whose dissolved oxygen content is predetermined.
- The whole solution is incubated in a closed bottle at 20°C for 5 days.
- After incubation the unused oxygen is determined.
- The difference between the original value of oxygen content in the diluted water and unused oxygen of solution after 5 days gives BOD.

5.7.2 Chemical oxygen demand (COD)

COD test is used to measure the organic matter of water and waste waters.COD is a measure of oxidisable sewage. It includes both the biologically oxidisable and biologically inert matter such as cellulose, as a result of which the value of COD is more than BOD. *COD is defined as the amount of oxygen (in ppm) consumed under specified conditions, while oxidizing total organic load of the sample with a strong oxidizing agent (Ex: potassium dichromate) in the acid medium.*

Determination COD

- A definite volume of waste water sample ('x' ml) is refluxed with a known volume of $K_2Cr_2O_7$ in H_2 SO_4 medium in the presence of $AgSO_4$ (which acts as a catalyst) and $HgSO_4$ (which eliminates interference due to chlorine).
- $K_2Cr_2O_7$ oxidises all organic matter into water, CO_2 and ammonia.
- The unreacted dichromate is titrated with a standard solution of ferrous ammonium sulphate (FAS) (Let the volume consumed is v_2 ml).
- COD = $\frac{(v_1 - v_2)N_{FAS} \times 8 \times 1000}{x}$; v_1 corresponds to the volume of FAS consumed in the blank titration (i.e., in the absence of waste water sample).

5.7.3 Dissolved Oxygen (DO)

DO is indicative of the amount of oxygen available for life to survive in an aquatic environment. DO is affected by the presence of metals, chemicals, nitrates, temperature fluctuations and organic life in the water. The less the DO in a body of water, the less favorable environment is for fish.

5.7.4 Total Organic Carbon (TOC)

TOC test is used for measuring small concentrations of organic matter in water. Injecting a small quantity of sample into a high-temperature furnace or chemically oxidizing environment performs the test.

5.7.5 pH

pH is an indicator of the acidity or alkalainity in a body of water. The concentration of carbonates and CO2 is the main influence on the pH of clean water. High concentration of bicarbonate produces alkaline waters (high pH), while low concentrations usually produce acidic waters (low pH)

Problems

1. Calculate the COD of the effluent sample when 25 ml of an effluent requires 8.3 ml of 0.001M $K_2Cr_2O_7$ for oxidation. [Given molecular mass of $K_2Cr_2O_7$=294).

Solution: Given Concentration of $K_2Cr_2O_7$=0.001M

Molecular mass of $K_2Cr_2O_7$=294

Volume of the effluent sample =25 ml

Volume of the $K_2Cr_2O_7$ consumed by the effluent =8.3ml

(i) 1000ml of 1M $K_2Cr_2O_7$=294 g

8.3 ml of 0.001M $K_2Cr_2O_7$= (294×8.3×0.001)/1000

Amount of $K_2Cr_2O_7$ present =2.4402mg

(ii) 1mol of $K_2Cr_2O_7$ ≡6 equivalents of oxygen

i.e., 294 mg of $K_2Cr_2O_7$ ≡6×8 mg of oxygen

$$\therefore 2.4402 \text{ mg of } K_2Cr_2O_7 \equiv \frac{6\times 8\times 2.4402}{294} = 0.3984 \text{ mg}$$

(iii) COD in 25 ml of water = 0.3984 mg

1000ml of water =398.4/25 =15.92 mg

$\therefore$ COD of water =15.92 mg/dm^3

2. What would be BOD value for a sample containing 200mg/dm^3 of glucose assuming that it was completely oxidized in the BOD test? (Atomic wt. of C = 12; H = 1; O = 16).

$$C_6H_{12}O_6 + 6O_2 \rightarrow 6CO_2 + 6H_2O$$

Molecular mass of glucose = 180 g

From the above equation, 180g of glucose requires 192 g of oxygen

$$\therefore 200 \text{ mg of glucose} \equiv \frac{192 \times 200}{180} = 213.33 \text{ mg}$$

Review Questions

1. Explain the urban problems related to water and how it can be tackled
2. Give the major water pollutants with example.
3. What is meant by point and nonpoint sources?
4. Define BOD and COD.
5. Enumerate with example the major sources of surface and ground water
6. Pollution.
7. Explain the causes, effects & control measures of Water pollution.
8. Explain the method of Industrial waste water treatment.
9. Explain the steps involved in domestic sewage water treatment.

CHAPTER 6

Soil Pollution

Soil Pollution – Soil Profile, Pollutants in soil, their adverse effects, controlling measures.

6.1 Soil

Introduction: is a mixture of abiotic factors that have broken down and settled into layers. Soil is the thin layer of organic and inorganic materials that covers the Earth's rocky surface.

Importance & Uses of Soil: Soil is also a place of intense biological activity thanks to degradation of organic matter, recycling of nutrients and synthesis of humus. The organic portion, which is derived from the decayed remains of plants and animals, is concentrated in the dark uppermost topsoil. The inorganic portion made up of rock fragments, was formed over thousands of years by physical and chemical weathering of bedrock. Productive soils are necessary for agriculture to supply the world with sufficient food. Fig. 6.1 illustrates the importance of Soil.

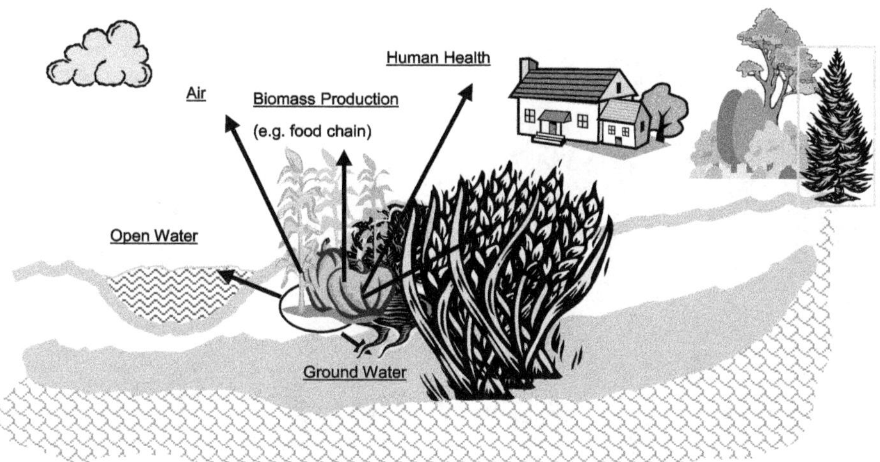

Fig. 6.1 Uses of Soil

6.2 Soil Profile

Soil profile refers to the layers of soil. Organic matter and other abiotic factors break down and settle into layers. A soil profile is a cross section through the soil which reveals its horizons (layers). Fig. 6.2 Shows Soil Profile.

There are six layers to the soil profile. They are as follows from the top:

O Horizon: (1st layer) This is the top layer of soil. Animals live on this layer. It is made of fresh to partially decomposed organic matters. The color varies from brown to black.

A Horizon**:** (2nd layer) The top part of this soil is made of highly decomposed organic matter mixed up. The color range from brown to gray.

E Horizon: (3rd layer) This layer is made up of mostly sand and silt it have lost most of its minerals and clay due to eluviation.

B Horizon**:** (4th layer) Unlike the other horizons, this one has more clay and bigger bedrock. It is reddish brown or tan in color.

C Horizon: (5th layer) This layer have mostly weathered bedrock. It is the cracked and broken surface of the bedrock.

R Horizon: (Last Layer) This is the last layer in the profile. It is made of unweathered rocks.

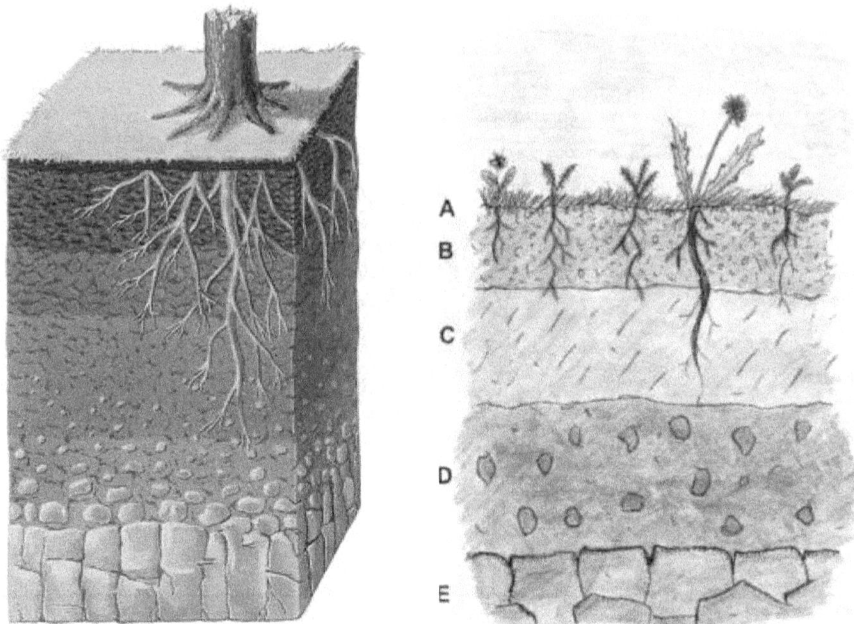

Fig. 6.2 Soil Profile

Soil generally consists of visually and texturally distinct layers, which can be summarized as follows, from top to bottom:

A. **Humus**: organic matter in relatively un decomposed form. This layer tends to be dark and rich in smell and texture.

B. **Topsoil**: well decomposed organic matter, mixed with a smaller amount of minerals.

C. **Layer** of mixed decomposed organic matter and mineral content.

D. **Subsoil** or mineral layers, the content of which varies according to the nature of the soil and its parent material.

E. **Bedrock** or parent material, which breaks down at the upper surface due to the effects of weathering and decay.

6.3 Land Pollution

is the contamination of the Earth's land surface through misuse of the soil by poor agricultural practices, mineral exploitation, industrial waste dumping, and indiscriminate disposal of urban wastes. It includes visible waste and litter as well as pollution of the soil itself.

These problems result in the loss of 6 million hectares of land each year. It also results in the loss of 24 billion tons of topsoil each year and a loss of at least 15 million acres of prime agricultural land to overuse and mismanagement every year

Soil Pollution: Soil often acts as the ultimate 'sink' of environmental pollution, because clay minerals and humic materials have a large number of surfaces, chemical groups and organic particles to which pollutants can attach. Contaminated soils can pose a problem for society if agricultural functions, human health or ecological systems are adversely affected.

Definition: Soil pollution is defined as the build-up in soils of persistent toxic compounds, chemicals, salts, radioactive materials, or disease causing agents, which have adverse effects on plant growth and animal health.

The introduction of substances, biological organisms, or energy into the soil, resulting in a change of the soil quality, which is likely to affect the normal use of the soil or endangering public health and the living environment.

6.4 Causes of Soil Pollution

Soil pollution is taking place due to many causes is caused by chemicals in herbicides and pesticides for agricultural activities as well as littering of waste materials in public places such as streets, parks and roads.. Soil pollution is caused by the presence of man-made chemicals or other alteration in the natural soil environment. This type of contamination typically arises from the rupture of underground storage links, application of pesticides, percolation of contaminated surface water to subsurface strata, oil and fuel dumping, leaching of wastes from landfills or direct discharge of industrial wastes to the soil. The most common chemicals involved are petroleum hydrocarbons, solvents, pesticides, lead and other heavy metals. This occurrence of this phenomenon is correlated with the degree of industrialization and intensities of chemical usage. A soil pollutant is any factor which deteriorates the quality, texture and mineral content of the soil or which disturbs the biological balance of the organisms in the soil. Pollution in soil has adverse effect on plant growth. There are many different ways that soil can become polluted, such as:

- Seepage from a landfill
- Discharge of industrial waste into the soil

- Percolation of contaminated water into the soil
- Rupture of underground storage tanks
- Excess application of pesticides, herbicides or fertilizer
- Solid waste seepage
- Unnecessary burning of forests,
- Surplus use of water,
- Unscientific land use etc

The most common chemicals involved in causing soil pollution are:

- Petroleum hydrocarbons
- Heavy metals
- Pesticides
- Solvents

Pollution in soil is associated with

- Indiscriminate use of fertilizers
- Indiscriminate use of pesticides, insecticides and herbicides
- Dumping of large quantities of solid waste
- Deforestation and soil erosion
- Pollution due to urbanisation
- Natural Land Pollution

(i) ***Indiscriminate use of fertilizers*** : Soil nutrients are important for plant growth and development. Plants obtain carbon, hydrogen and oxygen from air and water. But other necessary nutrients like nitrogen, phosphorus, potassium, calcium, magnesium, sulfur and more must be obtained from the soil. Farmers generally use fertilizers to correct soil deficiencies. Fertilizers contaminate the soil with impurities, which come from the raw materials used for their manufacture. Mixed fertilizers often contain ammonium nitrate (NH4NO3), phosphorus as P2O5, and potassium as K2O. For instance, As, Pb and Cd present in traces in rock phosphate mineral get transferred to super phosphate fertilizer. Since the metals are not degradable, their accumulation in the soil above their toxic levels due to excessive use of phosphate fertilizers becomes an indestructible poison for crops.

The over use of NPK fertilizers reduce quantity of vegetables and crops grown on soil over the years. It also reduces the protein content of wheat, maize, grams, etc., grown on that soil. The carbohydrate quality of such crops also gets degraded. Excess potassium content in soil decreases.

Vitamin C and carotene content in vegetables and fruits. The vegetables and fruits grown on over fertilized soil are more prone to attacks by insects and disease.

(ii) ***Indiscriminate use of pesticides, insecticides and herbicides*:** Plants on which we depend for food are under attack from insects, fungi, bacteria, viruses, rodents and other animals, and must compete with weeds for nutrients. To kill unwanted populations living in or on their crops, farmers use pesticides. The first widespread insecticide use began at the end of World War II and included DDT (dichlorodiphenyltrichloroethane) and gammaxene. Insects soon became resistant to DDT and as the chemical did not decompose readily, it persisted in the environment. Since it was soluble in fat rather than water, it biomagnified up the food chain and disrupted calcium metabolism in birds, causing eggshells to be thin and fragile. As a result, large birds of prey such as the brown pelican, ospreys, falcons and eagles became endangered. DDT has been now been banned in most western countries. Ironically many of them including USA, still produce DDT for export to other developing nations whose needs outweigh the problems caused by it.The most important pesticides are DDT, BHC, chlorinated hydrocarbons, organophosphates, aldrin, malathion, dieldrin, furodan, etc. The remnants of such pesticides used on pests may get adsorbed by the soil particles, which then contaminate root crops grown in that soil. The consumption of such crops causes the pesticides remnants to enter human biological systems, affecting them adversely.

An infamous herbicide used as a defoliant in the Vietnam War called Agent Orange (dioxin), was eventually banned. Soldiers' cancer cases, skin conditions and infertility have been linked to exposure to Agent Orange.

Pesticides not only bring toxic effect on human and animals but also decrease the fertility of the soil. Some of the pesticides are quite stable and their bio- degradation may take weeks and even months.

Pesticide problems such as resistance, resurgence, and heath effects have caused scientists to seek alternatives. Pheromones and hormones to attract or repel insects and using natural enemies or sterilization by radiation have been suggested.

(iii) ***Dumping of solid wastes*:** In general, solid waste includes garbage, domestic refuse and discarded solid materials such as those from commercial, industrial and agricultural operations. They contain increasing amounts of paper, cardboards, plastics, glass, old construction material, packaging material and toxic or otherwise hazardous substances. Since a significant amount of urban solid waste tends to be paper and food waste, the majority is recyclable or biodegradable in landfills. Similarly, most agricultural waste is recycled and mining waste is left on site.

The portion of solid waste that is hazardous such as oils, battery metals, heavy metals from smelting industries and organic solvents are the ones we have to pay particular attention to. These can in the long run, get deposited to the soils of the surrounding area and pollute them by altering their chemical and biological properties. They also contaminate drinking water aquifer sources. More than 90% of hazardous waste is produced by chemical, petroleum and metal-related industries and small businesses such as dry cleaners and gas stations contribute as well.

Solid Waste disposal was brought to the forefront of public attention by the notorious Love Canal case in USA in 1978. Toxic chemicals leached from oozing storage drums into the soil underneath homes, causing an unusually large number of birth defects, cancers and respiratory, nervous and kidney diseases.

(iv) ***Deforestation*****:** Soil Erosion occurs when the weathered soil particles are dislodged and carried away by wind or water. Deforestation, agricultural development, temperature extremes, precipitation including acid rain, and human activities contribute to this erosion. Humans speed up this process by construction, mining, cutting of timber, over cropping and overgrazing. It results in floods and cause soil erosion.

Forests and grasslands are an excellent binding material that keeps the soil intact and healthy. They support many habitats and ecosystems, which provide innumerable feeding pathways or food chains to all species. Their loss would threaten food chains and the survival of many species. During the past few years quite a lot of vast green land has been converted into deserts. The precious rain forest habitats of South America, tropical Asia and Africa are coming under pressure of population growth and development (especially timber, construction and agriculture). Many scientists believe that a wealth of medicinal substances including a cure for cancer and aids, lie in these forests. Deforestation is slowly destroying the most productive flora and fauna areas in the world, which also form vast tracts of a very valuable sink for CO_2.

(v) ***Pollution Due to Urbanisation*:** Urban activities generate large quantities of city wastes including several Biodegradable materials (like vegetables, animal wastes, papers, wooden pieces, carcasses, plant twigs, leaves, cloth wastes as well as sweepings) and many non-biodegradable materials (such as plastic bags, plastic bottles, plastic wastes, glass bottles, glass pieces, stone / cement pieces).

On a rough estimate Indian cities are producing solid city wastes to the tune of 50,000 - 80,000 metric tons every day. If left uncollected and decomposed, they are a cause of several problems such as

- *Clogging of drains*: Causing serious drainage problems including the burst / leakage of drainage lines leading to health problems.
- *Barrier to movement of water*: Solid wastes have seriously damaged the normal movement of water thus creating problem of inundation, damage to foundation of buildings as well as public health hazards.

- *Foul smell*: Generated by dumping the wastes at a place.
- *Increased microbial activities*: Microbial decomposition of organic wastes generate large quantities of methane besides many chemicals to pollute the soil and water flowing on its surface.
- When such solid wastes are hospital wastes they create many health problems: As they may have dangerous pathogen within them besides dangerous medicines, injections.
- Underground soil in cities is likely to be polluted by.
- Chemicals released by industrial wastes and industrial wastes.
- Decomposed and partially decomposed materials of sanitary wastes.

Many dangerous chemicals like cadmium, chromium, lead, arsenic, selenium products are likely to be deposited in underground soil. Similarly underground soil polluted by sanitary wastes generates many harmful chemicals. These can damage the normal activities and ecological balance in the underground soil.

(vi) ***Natural land pollution*:** Land pollution occurs massively during earth quakes, land slides, hurricanes and floods. All cause hard to clean mess, which is expensive to clean, and may sometimes take years to restore the affected area. These kinds of natural disasters are not only a problem in that they cause pollution but also because they leave many victims homeless.

(vii) **Causes of Soil Pollution in brief:**

- Polluted water discharged from factories
- Runoff from pollutants (paint, chemicals, rotting organic material) leaching out of landfill
- Oil and petroleum leaks from vehicles washed off the road by the rain into the surrounding habitat

- Chemical fertilizer runoff from farms and crops
- Acid rain (fumes from factories mixing with rain)
- Sewage discharged into rivers instead of being treated properly
- Over application of pesticides and fertilizers
- Purposeful injection into groundwater as a disposal method
- Interconnections between aquifers during drilling (poor technique)
- Septic tank seepage
- Lagoon seepage
- Sanitary/hazardous landfill seepage
- Cemeteries
- Scrap yards (waste oil and chemical drainage)
- Leaks from sanitary sewers

6.5 Types of Soil Pollution

Agricultural Soil Pollution

(i) pollution of surface soil

(ii) pollution of underground soil

Soil pollution by industrial effluents and solid wastes

(i) pollution of surface soil

(ii) disturbances in soil profile

Pollution due to urban activities

(i) pollution of surface soil

(ii) pollution of underground soil

6.6 Effects of Soil Pollution

When it comes to the environment itself, the toll of contaminated soil is even more dire. Soil that has been contaminated should no longer be used to grow food, because the chemicals can leech into the food and harm people who eat it. If contaminated soil is used to grow food, the land will usually produce lower yields than it would if it were not contaminated. This, in turn, can cause even more harm because a lack of plants on the soil will cause more erosion, spreading the contaminants onto land that might not have been tainted before. In addition, the pollutants will change the makeup of the soil and the types of microorganisms that will live in it. If certain organisms die off in the area, the larger predator animals will also have to move away or die because they've lost their food supply. Thus it's possible for soil pollution to change whole ecosystems

6.6.1 Agricultural

- Reduced soil fertility
- Reduced nitrogen fixation
- Increased erodibility
- Larger loss of soil and nutrients
- Deposition of silt in tanks and reservoirs
- Reduced crop yield
- Imbalance in soil fauna and flora

6.6.2 Industrial

- Dangerous chemicals entering underground water
- Ecological imbalance
- Release of pollutant gases
- Release of radioactive rays causing health problems
- Increased salinity
- Reduced vegetation

6.6.3 Urban

- Clogging of drains
- Inundation of areas
- Public health problems
- Pollution of drinking water sources
- Foul smell and release of gases
- Waste management problems

6.6.4 Effects of soil pollution in brief

- pollution runs off into rivers and kills the fish, plants and other aquatic life

- crops and fodder grown on polluted soil may pass the pollutants on to the consumers
- polluted soil may no longer grow crops and fodder
- Soil structure is damaged (clay ionic structure impaired)
- corrosion of foundations and pipelines
- impairs soil stability
- may release vapours and hydrocarbon into buildings and cellars
- may create toxic dusts
- may poison children playing in the area

6.7 Control of Soil Pollution

The following steps have been suggested to control soil pollution.

- Minimising the usage of pesticides.
- Periodic change of crops to increase the soil fertility.
- Disposal of unwanted garbage created by clinics and hospitals properly either by burning or by burying into the soil.
- Following the rules and regulations laid down by pollution control board.
- Minimising the usage of plastics.

(i) ***Reducing chemical fertilizer and pesticide use*****:** Applying bio-fertilizers and manures can reduce chemical fertilizer and pesticide use. Biological methods of pest control can also reduce the use of pesticides and thereby minimize soil pollution.

(ii) ***Reusing of materials*****:** Materials such as glass containers, plastic bags, paper, cloth etc. can be reused at domestic levels rather than being disposed, reducing solid waste pollution.

(iii) ***Recycling and recovery of materials*****:** This is a reasonable solution for reducing soil pollution. Materials such as paper, some kinds of plastics and glass can and are being recycled. This decreases the volume of refuse and helps in the conservation of natural resources. For example, recovery of one tonne of paper can save 17 trees.

(iv) ***Reforesting*****:** Control of land loss and soil erosion can be attempted through restoring forest and grass cover to check wastelands, soil erosion and floods. Crop rotation or mixed cropping can improve the fertility of the land.

(v) ***Solid waste treatment*:** Proper methods should be adopted for management of solid waste disposal. Industrial wastes can be treated physically, chemically and biologically until they are less hazardous. Acidic and alkaline wastes should be first neutralized; the insoluble material if biodegradable should be allowed to degrade under controlled conditions before being disposed.As a last resort, new areas for storage of hazardous waste should be investigated such as deep well injection and more secure landfills. Burying the waste in locations situated away from residential areas is the simplest and most widely used technique of solid waste management. Environmental and aesthetic considerations must be taken into consideration before selecting the dumping sites. Incineration of other wastes is expensive and leaves a huge residue and adds to air pollution. Pyrolysis is a process of combustion in absence of oxygen or the material burnt under controlled atmosphere of oxygen. It is an alternative to incineration. The gas and liquid thus obtained can be usedas fuels. Pyrolysis of carbonaceous wastes like firewood, coconut, palm waste, corn combs, cashew shell, rice husk paddy straw and saw dust, yields charcoal along with products like tar, methyl alcohol acetic acid, acetone and a fuel gas.

(vi) ***Prevention of soil degradation & Erosion*:** Soil erosion can be defined as the movement of surface litter and topsoil from one place to another. While erosion is a natural process, often caused by wind and flowing water, it is greatly accelerated by human activities such as farming, construction, overgrazing by livestock, burning of grass cover, and deforestation.

The loss of the topsoil makes a soilless fertile and reduces its water-holding capacity. The topsoil, which is washed away, also contributes to water pollution by clogging lakes and increasing the turbidity of the water, ultimately leading to the loss of aquatic life. For one inch of topsoil to be formed it normally requires 200-1000 years, depending upon the climate and soil type. Thus, if the topsoil erodes faster than it is formed, the soil becomes a non-renewable resource. Therefore, it is essential that proper soil conservation measures are, used to minimize the loss of the topsoil. There are several techniques that can protect the soil from erosion.

Today, both water and soil are conserved through integrated treatment methods. The two types of treatment generally used are:

- Area treatment, which involves treating the land
- Drainage-line treatment, which involves treating the natural water.

(a) Area Treatment

Purpose	Treatment Measure	Effect
Reduces the impact of rain drops on the soil	Develop vegetative cover on the non arable land	Minimum disturbance and displacement of soil particles
Infiltration of water where it falls	Apply water infiltration measures on the area	In-situ soil and moisture conservation
Minimum surface run-off	Store surplus rain water by constructing bunds, ponds in the area	Increased soil moisture in the area, facilitate ground water recharge
Ridge to valley sequencing	Treat the upper catchment first and then proceed towards the outlet	Economically viable, less riskof damage and longer life of structures of the lower catchments

(b) Drainage-line treatment

Purpose	Treatment Measure	Effect
Stop further deepening of gullies and retain sediment run-off	Plug the gullies at formation	Stops erosion, recharges groundwater at the upper level
Reduce run-off velocity, pass cleaner water to the downstream side	Create temporary barriers in nalas	Delayed flow and increased groundwater recharge
Minimum sedimentation in the storage basins	Use various methods to treat the catchments	
Low construction cost	Use local material and skills for constructing the structures	Structures are locally maintained

Continuous contour trenches can be used to enhance the infiltration of water, reduce the run-off, and check soil erosion. These are actually shallow trenclles dug across the slope of the land and along the contour lines, basically for the purpose of soil and water conservation. They are most effective on gentle slopes and in areas of low to medium rainfall. These bunds are stabilized by fast-growing tree species and grasses. In areas with steep slopes where bunds are not possible, continuous contour benches (CCBs) made of stones are used for the same purpose. Gradonies can also be used to convert wastelands into agricultural lands. In this, narrow trenches with bunds on the downstream side are built along contours in the upper reaches of the catchment to collect run-off and to conserve moisture from the trees or tree crops. The area between the two bunds is used for cultivating crops after development of fertile soil cover.

Some of the ways in which this can be achieved are:

- Live check-dams, in which barriers are created by planting grass, shrubs and trees across the gullies.
- A bund constructed out of stones across the stream can also be used for conserving soil and water.
- An earthen check-bund constructed out of local soil across the stream to check soil erosion and the flow of water.
- A Gabion structure, which is a bund constructed of stone and wrapped in galvanized chain link. A gabion structure has a one-inch thick, impervious wall of ferrocement at the center of the structure, which goes below the ground level up to the hard strata. This ferrocement partition, supported by the gabion portion, is able to retain the water and withstand the force of the runoff water.
- An bandhara is an underground structure across a nala bed that functions as a barrier to check the movement of groundwater.

6.8 Soil Pollution Sources, Effects & Prevention in Brief

- ***Sources:*** oil spills; sewage and waste dumping; the mishandling of solid waste, which is garbage; deforestations; pesticides and use of other chemicals; deforestation as that reduces the amount of oxygen produced...
- ***Effects***: desertification, which means that good cultivation land can turn into deserts; a decrease in crops; wildlife becoming extinct or dying…
- ***Prevention:*** Education of people through campaigns, planting crops, or preserving the animals and wildlife, recycling to reduce wastes. Stringent Policies and legislation by government

Review Questions

1. What are the major sources of soil pollution?
2. What are the measures to be taken to prevent soil pollution?
3. What are the causes of soil erosion and methods of preventing it.
4. Define soil & land pollution.
5. Explain the sources, causes, effects and control measures of Soil pollution

CHAPTER 7

Society and Ecology

Impact of waste on society. Solid waste management (Nuclear, Thermal, Plastic, medical, Agriculture, domestic and e-waste).

7.1 Impact of Waste on Society

All living beings including man are dependant on their environment for existence. But every manmade activity has some impact on the environment. More often it is harmful than benign. But human beings cannot live without taking up these activities for their food, shelter, comfort, security and many other needs. The following activities cause major impacts on the environment: Agriculture, Housing, Transportation, Industries, Water resources projects including irrigation projects, Power Generation, Mining, Tourism, Socio-Economic activities, Defense related activities, Petroleum processing, Urbanization, Commercial deforestation, Providing public amenities such as water supply, sanitation, electricity, telephone, transportation etc.,Religious places – public activities

Every activity of the man from birth to death has its impact on the environment. Some of the major impacts are listed below:

Activity	Impacts
Agriculture	• Soil erosion • Discharge of nutrients into water bodies /ground water • Discharge of pesticides into the environment these pesticides end up in the food chain of the ecosystem. • Endosulfan problem of cashew nut farms in Kerala which has crippled human beings is a living example. • Imposing Water burden on water resources • Water pollution
Water Resources projects	• Deforestation • Submergence of forest and other lands • Water logging problems • Evacuation and rehabilitation of people and villages • Disturbance to wild life • Mosquito breeding
Housing	• Extraction of construction material • Cutting of forests • Energy utilization • Stress on water resources • Urban centers impose heavy burden on the environment • Disruption of storm water drainage pattern
Transportation	• Deforestation for constructing roads and railways • Utilization of valuable agricultural land for construction of airports which change the land use pattern • Air pollution • Noise pollution • Disruption of wild life habitats • Pollution of marine waters due to harbours

Industries	• Pressure on land and other natural resources for raw material • Water pollution • Air pollution • Noise pollution • Pressure on transport systems
Power Generation	• Hydroelectric plants–submergence of valuable lands, deforestation, disruption of wild life etc., • Thermal power plants create water pollution, air pollution and thermal pollution problem besides requiring large quantity of water • Power transmission lines lead to deforestation • Thermal power plants require coal. • Coal mining is environmentally critical activity. Also coal has to be hauled over long distances creating transportation related problems. • Nuclear power plants carry the risk of radioactive hazards. • Global warming / climate change and acid rain are related to combustion of fossil fuels in thermal power plants.
Mining	• Deforestation • Large tracts of land is made barren • Air pollution • Water pollution • Soil erosion • Transportation of ores imposes heavy burden on transport facilities
Tourism and Religious activities	• Create congestion • Transport problems • Sanitation problems • Water supply related problems • Spread of diseases

- Social problems
- Accumulation of plastic and other solid wastes

Human Habitation and Urbanization

- Growth of urban centres create all sorts of environmental problems like air, water and noise pollution, traffic related problems sanitation problems etc.
- Solid waste generation
- Water burden
- Social tensions

7.2 Solid Waste Management

Rapid population growth and urbanization developing countries have led to the generation of enormous quantities of solid waste and consequent environmental degradation. An estimated 7.6 million tones of municipal solid wastes are disposed in open dumps creating considerable nuisance and environmental problems.

7.2.1 Types/ Classification of Solid Waste

Solid waste can be classified in many different ways, but the following classification is generally found based on generation of Solid waste. They are:

(i) ***Domestic waste*****:** This includes both urban and rural waste. The most important characteristics of these waste is that they are highly putrescible and will decompose rapidly, especially in warm weather. Often decomposition will lead to the development of offensive odors.

(ii) ***Industrial waste/ Treatment Plants waste*****:** The solid and semi solid waste from water, waste water and industrial waste water characteristics of these materials vary depending on the nature of the treatment process. The main categories of industrial waste generated includes mining sterile water, ashes and slag from thermal plants, metallurgy waste, residual sludge, chemical waste, and ferrous waste.

(iii) ***Agricultural waste*****:** Waste is resulting from various agricultural activities such as planting and harvesting of crops.

(iv) ***Hazardous waste:*** This includes chemical, biological, flammable, explosive or radiological wastes that pose a substantial danger to human, animal and plant life.

(v) ***Demolition and Construction waste***: Wastes from the construction, remodelling and repair of individual residences, commercial buildings, and repairing of individual residences, construction waste. They include stones, bricks, concrete, plaster dirt, and electrical parts.

7.2.2 Treatment Methods/Strategy

The strategy used to develop an integrated waste management system is to identify the level or levels at which the highest values of individual and collective materials can be recovered. For this reason, the list starts with reduction using less and reusing more, thereby saving material production, resource cost, and energy. At the bottom of the list is ultimate disposal the final resting place for waste (Fig. 7.1).

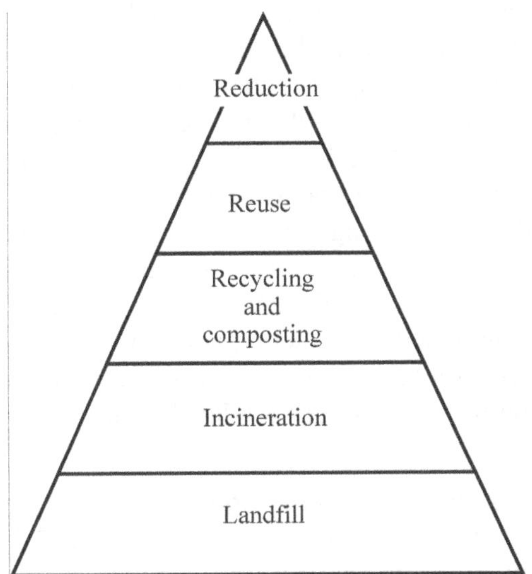

Fig. 7.1 Hierarchy of solid waste management.

(i) ***Reduction Strategies:*** Reduction strategies are any approaches a community may use to lower the amount of waste being produced. Some simple reduction activities that individuals within a community can do are backyard composting (this reduces the amount of waste disposed of in landfills) and two-sided copying on paper. A waste exchange program also contributes to reduction. In any of the change strategies (reduction, reuse, and recovery), public education and involvement are crucial, and in the case of reduction, they are imperative. Reduction assumes the commitment and involvement of all citizens. Source reduction strategies have many favorable environmental impacts, including reducing greenhouse gas production, saving energy, and conserving resources, in addition to reducing the volume of the waste stream.

(ii) ***Reuse Strategies:*** Reuse is using a product more than once, either for the same purpose or for an alternate purpose. Reuse does not require reprocessing and, therefore, has lower energy requirements than recycling. Reuse strategies include making donations to charity, reusing packaging (including boxes and bags), using empty jars for food storage, and participating in a paint collection and reuse program.

(iii) ***Materials Recovery, Recycling and Composting:*** In recycling, waste materials are processed industrially and then reformed into new or similar products. Recycling includes preconsumer waste, such as factory cuttings or shavings, as well as post-consumer waste items, including cardboard, newspapers, plastic bottles, and aluminum cans. Although recycling is often viewed as a resource conservation activity, it may offer greater return for many products in terms of energy savings. A second means of recapturing value is through the use of the natural biodegradation process. In urban areas, the composting of leaf and tree waste alone can reduce landfill dependency. The segregation of yard waste from other organic (biodegradable) wastes is necessary to avoid contamination of the compost which might render the mulch or end product less desirable.

(iv) ***Resource Recovery & Incineration:*** The third approach to value recapture is to incinerate waste and use the heat for energy. Although many combustibles are recyclable, there is often a higher total value (due to processing costs) in burning the waste for energy than in recycling. Often, many combustible/recyclable materials are contaminated and rendered difficult and/or expensive to recycle. By developing an incineration program with a materials recovery component, furnace and processing equipment life is usually extended because glass and ferrous and non-ferrous metals are removed during material recovery. Incineration reduces the volume of refuse by up to 90 percent, leaving behind only ash, and resulting in less need for landfill space.

(v) ***Ultimate Disposal & Landfill:*** The last option is disposal. Given current technology, there are residuals from the previous processes, and some materials are simply not recoverable and must go somewhere. The continuing development of more stringent requirements for landfills is making this ultimate disposal option less environmentally offensive, but more costly. An integrated waste management system entails a careful analysis of what is in the waste stream and offers ideas on practices to recover the various materials at the point of highest value. The best strategy for a community is to match its unique position with the mix of activities that will best serve it now and far into the future.

7.2.3 Characterisation of Wastes

A solid waste is a hazardous waste if it exhibits any of the characteristics identified here in below:

(i) ***Ignitability:*** A solid waste exhibits a characteristic of ignitability if a representative sample of the waste has any of the following properties:

S.No.	Form	Particulars
1.	Liquid (other than an aqueous solution containing less than 24% alcohol by volume)	has a flash point less than 60 degrees Celsius
2.	Not a liquid	Under standard temperature and pressure, capable of causing fire through friction, absorption of moisture or spontaniouschemical changes When ignited, burns vigorously and persistently
3.	Compressed gas	If ignitable
4.	Solid Wastes	If ignitable
5.	Oxidiser	

(ii) ***Corrosivity:*** A solid waste exhibits a characteristic of corrosivity if a representative sample of the waste has either of the following properties:

S. No.	Form	Particulars
1.	Aqueous	Has a pH less than 2 or greater than 12.5
2.	Liquid	Capable to corrode steel at a rate greater than 6.35 mm per year at a temperature 55 degrees

(iii) ***Characteristic of Reactivity:*** A solid waste exhibits a characteristic of reactivity if a representative sample of the waste has any of the following properties: It is normally unstable and readily undergoes violent change without detonating. It reacts violently with water. It forms potentially explosive mixture with water. When mixed with water, it generates toxic gases, vapors or fumes in a quantity sufficient to present a danger to human health or the environment. It is a cyanide or sulfide bearing waste which, when exposed to pH conditions between 2 and 12.5, can generate toxic gases, vapors or fumes in a quantity sufficient to present a danger to human health or the environment. It is capable of detonation or explosive reaction if it is subjected to a strong initiating source or if heated under confinement. It is readily capable of detonation or explosive decomposition of reaction at standard temperature and pressure. It is a forbidden explosive.

(iv) ***Toxicity Characteristics*****:** A solid waste exhibits the characteristics of toxicity if using the Toxicity Characteristic Leaching procedure, the extract from a representative sample of the waste contains any of the concentration equal to or greater than the respective value given in the Table 7.1. Where the waste contains less than 0.5% filterable solids, the waste itself, after filtering is considered to be extract for the purpose of this section.

Table 7.1 Maximum Concentration of contaminants for the Toxicity Characteristic

Arsenic	5.0	Hexachlorobenzene	0.13
Barium	100.0	Hexachlorobutadine	0.5
Benzene	0.5	Hexachloroethane	3.0
Cadmium	1.0	Lead	5.0
Carbom tetrachloride	0.5	Lindane	0.4
Chlordane	0.03	Mercury	0.2
Chloro Benzene	100.0	Methoxychlor	0.0
Chloroform	6.0	Methyl ethyl ketone	200.0
Chromium	5.0	Nitro benzene	2.0
O-Cresol	200.0	Pentrachlorophenol	100.0
m-Chesol	200.0	Pyridine	5.0
p-Cresol	200.0	Selenium	1.0
Cresol	200.0	Silver	5.0
2,4-D	10.0	Tetrachloroethylene	0.7
1,4-Dichloro benzene	7.5	Toxaphene	0.5
1,2-Dichloro ethane	0.5	Trichloroethylene	0.5
1, 1 -Dichloroethylene	0.7	2,4,5-	400.0
2,4-Dichlorotoluene	0.13	Trichlorophenol	2.0
Endrin	0.02	2,4,6-	1.0
Heptachlor (and its..	0.008	Trichlorophenol	0.2
oxide		2,4,5-TP (Silver)	
		Vinyl Chloride	

Likewise if certain waste constituents are higher in the waste than a limit, it be segregated and not allowed to be disposed it by land-filling.

7.3 Nuclear or Radioactive Wastes

Radio active waste comprises a variety of materials requiring different types of management to protect people and the environment. Another factor in managing wastes is the time that they are likely to remain hazardous. This depends on the kinds of radioactive isotopes the half lives characteristic of each of those isotopes. The half life is the time it takes for a given radioactive isotope to lose half of its radioactivity. They are normally classified as low-level, medium-level or high-level wastes, according to the amount and types of radioactivity in them.

Low-level Waste: is generated from hospitals, laboratories and industry, as well as the nuclear fuel cycle. It comprises paper, rags, tools, clothing, filters etc. which contain small amounts of short-lived radioactivity. Usually it is buried in shallow landfill sites. To reduce its volume, it is often compacted or incinerated (in a closed container) before disposal. Worldwide it comprises 90% of the volume but only 1% of the radioactivity of all rad waste.

Intermediate-level Waste: Contains higher amounts of radioactivity and may require special shielding. It typically comprises resins, chemical sludges and reactor components, as well as contaminated materials from reactor decommissioning. Worldwide it makes up 7% of the volume and has 4% of the radioactivity of all radwaste. It may be solidified in concrete or bitumen for disposal. Generally short-lived waste (mainly from reactors) is buried, but long-lived waste (from reprocessing nuclear fuel) will be disposed of deep underground.

High-level Waste: May be the used fuel itself, or the principal waste from reprocessing this. While only 3% of the volume of all radwaste, it holds 95% of the radioactivity. It contains the highly-radioactive fission products and some heavy elements with long-lived radioactivity. It generates a considerable amount of heat and requires cooling, as well as special shielding during handling and transport. If the used fuel is reprocessed, the separated waste is vitrified by incorporating it into borosilicate (Pyrex) glass which is sealed inside stainless steel canisters for eventual disposal deep underground.

Three general principles are employed in the management of radioactive wastes:

concentrate-and-contain

dilute-and-disperse

delay-and-decay.

As a result of the operation of nuclear reactors, some radioactive wastes are produced. Yet compared to the amount of waste produced by coal-fired electrical generating plants, these are of considerably smaller volume. The wastes generated at nuclear power plants are rather low in activity and the radio nuclides contained therein have a low radio toxicity and usually a short half-life. However, nuclear power plants are the largest in number among all nuclear facilities and produce the greatest volume of radioactive wastes. The nature and amounts of wastes produced in a nuclear power plant depend on the type of reactor, its specific design features, its operating conditions and on the fuel integrity. These radioactive wastes contain activated radio nuclides from structural, moderator, and coolant materials; corrosion products; and fission product contamination arising from the fuel. The methods applied for the treatment and conditioning of waste generated at nuclear power plants now have reached a high degree of effectivity and reliability and are being further developed to improve safety and economy of the whole waste management system.

7.3.1 Wastes Generated at Nuclear Power Plants

Low- and intermediate-level radioactive waste (LILW) at nuclear power plants is produced by contamination of various materials with the radio nuclides generated by fission and activation in the reactor or released from the fuel or cladding surfaces. The radio nuclides are primarily released and collected in the reactor coolant system and, to a lesser extent, in the spent fuel storage pool. The main wastes arising during the operation of a nuclear power plant are components which are removed during refueling or maintenance (mainly activated solids, e.g. stainless steel containing cobalt-60 and nickel-63) or operational wastes such as radioactive liquids, filters, and ion-exchange resins which are contaminated with fission products from circuits containing liquid coolant.

In order to reduce the quantities of waste for interim storage and to minimize disposal cost, all countries are pursuing or intend to implement measures to reduce the volume of waste arisings where practicable. Volume reduction is particularly attractive for low-level waste which is generally of high volume but low radiation activity. Significant improvements can be made through administrative measures, e.g. replacement of paper towels by hot air driers, introduction of reusable long-lasting protective clothing, etc., and through general improvements of operational implementation or "housekeeping".

Liquid wastes and wet solid wastes: According to the different types of reactors now operating commercially all over the world, different waste streams arise. These streams are different both in activity content and in the amount of liquid waste generated. Reactors cooled and moderated by water generate more liquid waste than those cooled by gas. The volumes of liquid waste generated at boiling-water reactors (BWRs) are significantly higher than at pressurized water reactors (PWRs). Because the cleanup system of heavy-water reactors (HWRs) works mainly with once through ion-exchange techniques to recycle heavy water, virtually no liquid concentrates are generated at them.

Active liquid wastes are generated by the cleanup of primary coolants (PWR, BWR), cleanup of the spent fuel storage pond, drains, wash water, and leakage waters. Decontamination operations at reactors also generate liquid wastes resulting from maintenance activities on plant piping and equipment. Decontamination wastes can include crud (corrosion products) and a wide variety of organics, such as oxalic and citric acids.

Wet solids are another category of waste generated at nuclear power plants. They include different kinds of spent ion-exchange resins, filter media, and sludges. Spent resins constitute the most significant fraction of the wet solid waste produced at power reactors. Bead resins are used in deep demineralizers and are common in nuclear power plants. Powdered resins are seldom used in PWRs, but are commonly used in BWRs with pre-coated filter demineralizers. In many BWRs, a large source of powdered resin wastes are the "condensate polishers" used for additional cleaning of condensed water after evaporation of liquid wastes.

Pre-coated filters used at nuclear power plants to process liquid waste produce another type of wet solid waste-filter sludges. The filter aids — usually diatomaccous earth or cellulose fibers -and the crud that is removed from the liquid waste together form the filter sludges. Some filtration systems do not require filter aid materials. The sludges arising from such units therefore do not contain other materials.

7.3.2 Treatment and Conditioning of Liquid/solid Waste

Liquid radioactive waste generated at nuclear power plants usually contains soluble and insoluble radioactive components (fission and corrosion products) and non-radioactive substances. The general objective of waste treatment methods is to decontaminate liquid waste to nearly all processes are applied to treat radioactive effluents. Standard techniques are routinely used to decontaminate liquid waste streams. Each process has a particular effect on the radioactive content of the liquid. The extent to which these are used in combination depends on the amount and source of contamination. Four main technical processes are available for treatment of liquid waste: evaporation; chemical precipitation/flocculation; solid-phase separation; and ion exchange.

These treatment techniques are well established and widely used. Nevertheless, efforts to improve safety and economy on the basis of new technologies are under way in many countries.

The best volume reduction effect, compared with the other techniques, is achieved by evaporation. Depending on the composition of the liquid effluents and the types of evaporators, decontamination factors between 10^4 and 10^6 are obtained.

Evaporation is a proven method for the treatment of liquid radioactive waste providing both good decontamination and volume reduction. Water is removed in the vapour phase of the process leaving behind non-volatile components such as salts containing most radio nuclides. Evaporation is probably the best technique for wastes having relatively high salt content with a wide heterogenous chemical composition. Such an extent that the decontaminated bulk volume of aqueous waste can be either released to the environment or recycled. Waste concentrate is subject to further conditioning, storage, and disposal. Because nuclear power plants generate almost all categories of liquid waste.

7.4 Plastic Waste

Plastic waste commands the highest rate in the recycled market. Carrybags manufactured using lower grade recycled materials are unacceptable and are the main environmental culprits. The issue boils down to management of plastic waste, and more precisely carrybags and containers made out of recycled plastic waste material. It is said that any strategy for effective management of plastic wastes should have three R's - reduction, reuse, and recycle, and include a package of prevention, promotion, and mitigation measures. The more that is recycled, the longer will natural resources be available for future generations.An interesting economic fact is that recycled polythene bags are generally priced between Rs. 45 and 50 per kg, while bags made out of virgin plastics command a price of around Rs. 80 per kg. The disposables which generate waste and cause environmental problems when their useful life ends, include mainly the following:

- Plastic packaging/carrybags/bottles/containers/trash bags
- Plastics from health and medicare
- Plastics from hotels and catering industry
- Plastics from air, rail and road travel

The Union Ministry of Environment and Forests has recently notified the "Recycled Plastic Manufacture and Usage Rules, 1999". These rules require that carrybags or containers used for purposes of storing shall be made of virgin plastic and be in natural shade or white. These items when made of recycled plastic, and used for purposes other than storing and packaging of foodstuffs shall use pigments and colourants as per Indian Standards. Recycling of plastics shall also be undertaken strictly in accordance with specifications prescribed by the Bureau of Indian Standards, and shall carry a mark that the product is manufactured out of recycled plastic. The thickness of carrybags shall not be less than 20 microns. Finally and most importantly, Rule 4 prohibits all vendors from using carrybags or containers made out of recycled plastics for storing, carrying, dispensing or packaging of foodstuffs. In other words all vendors are required to use carrybags and containers manufactured to specifications prescribed in the 1999 Rules. Packaging is found wherever products are sold, be they foodstuffs or consumer goods. Recycling capacities, catering for plastic packaging of sweet wrappers to 10-litre mayonnaise buckets, are available with suitable processes for each type of packaging requirement. The recycling follows sound ecological routes which meet all statutory provisions and, at the same time remain within reasonable financial limits.

7.5 Medical Waste

Medical Waste" is defined as all waste generated in direct patient care or in diagnostic or research areas that is non-infectious but aesthetically repugnant if found in the environment. Bio-medical waste is defined as waste that is generated during the diagnosis, treatment or immunisation of human beings and are contaminated with patients' body fluids (such as syringes, needles, ampoules, organs and body parts, placenta, dressings, disposables plastics and microbiological wastes). Proper disposal of Hospital waste is of paramount importance because of its infectious and hazardous characteristics.

Human blood and blood products are classified and managed as medical waste because of the possible presence of infectious agents that cause blood-borne disease. Wastes in this category include bulk blood and blood products as well as smaller quantities of blood samples drawn for testing or research.

The pollutants from medical waste incinerators either exist in the waste feed material or are formed in the combustion process. There are organic emissions other than dioxins and furans that are generated by medical waste incineration, but their emission rates are usually not regulated.

Incineration of medical wastes with air pollution control equipment is required to meet new regulations.

Plasma pyrolysis provides solutions for complete pyrolysis of typical hospital waste such as cellulose polymer dressings, polyvinyl chloride blood bags, polyurethane and silicon rubber gloves & catheters and other disposables made of polyethylene, polymethyl methacrylate, rubber, glass etc.

7.5.1 Technology for Medical Waste Treatment

- Plasma Pyrolysis for Medical Waste
- Microwave Incineration to Process Biological Wastes
- Hypodermic Needle Medical Waste Disposal
- Microwave Reduction of Medical Waste
- Waste Minimization, Segregation and Recycling in Hospitals
- Dental waste recycling & disposal

7.6 Domestic Waste: Municipal Solid Waste (MSW)

Also called urban solid waste, is a waste type that includes predominantly household waste (domestic waste) with sometimes the addition of commercial wastes collected by a municipality within a given area. They are in either solid or semisolid form and generally exclude industrial hazardous wastes. The term *residual waste* relates to waste left from household sources containing materials that have not been separated out or sent for reprocessing.

Waste handling and separation involves the activities associated with management of waste until they are placed in storage container for collection. Handling also encompasses the movement of loaded containers to the point of collection. Separation of waste components is an important step in the handling and storage of solid waste at the source.

7.6.1 Transfer and Transport

This element involves two steps as given below.

(i) the transfer of wastes from the smaller collection vehicle to the larger transport equipment

(ii) the subsequent transport of the wastes, usually over long distances, to a processing or disposal site.

7.6.2 Disposal

Today the disposal of wastes by land filling or land spreading is the ultimate fate of all solid wastes, whether they are residential wastes collected and transported directly to a landfill site, residual materials from materials recovery facilities (MRFs), residue from the combustion of solid waste, compost or other substances from various solid waste processing facilities. A modern sanitary landfill is not a dump; it is an engineered facility used for disposing of solid wastes on land without creating nuisances or hazards to public health or safety, such as the breeding of insects and the contamination of ground water.

7.6.3 Energy Generation

Municipal solid waste can be used to generate energy. Several technologies have been developed that make the processing of MSW for energy generation cleaner and more economical than ever before, including landfill gas capture, combustion, pyrolysis, gasification, and plasma arc gasification. While older waste incineration plants emitted high levels of pollutants, recent regulatory changes and new technologies have significantly reduced this concern. EPA regulations in 1995 and 2000 under the Clean Air Act have succeeded in reducing emissions of dioxins from waste-to-energy facilities by more than 99 percent below 1990 levels, while mercury emissions have been reduced by over 90 percent The EPA noted these improvements in 2003, citing waste-to-energy as a power source "with less environmental impact than almost any other source of electricity." Fig. 7.2 shows waste to energy system

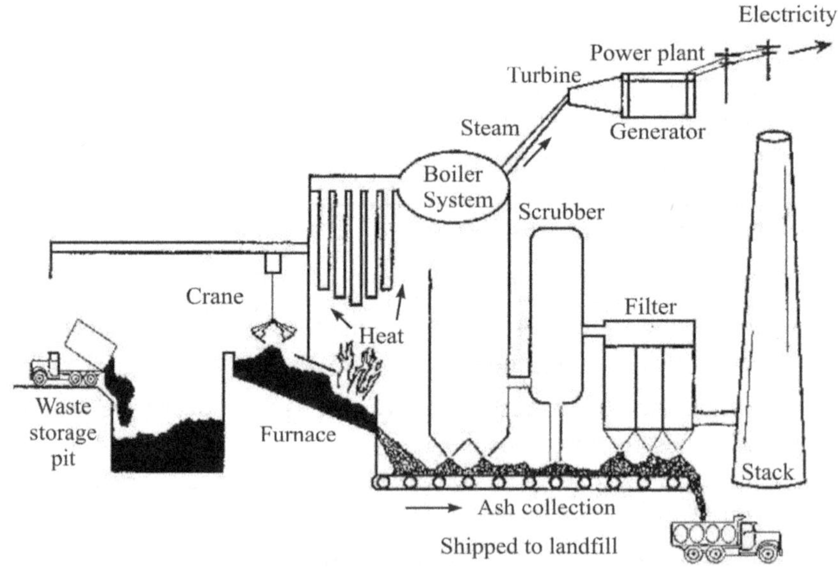

Fig. 7.2 Waste to Energy

7.7 E-Waste

Definition of E-waste: Electronic waste, e-scrap, or Waste Electrical and Electronic Equipment (WEEE) describe loosely discarded, surplus, obsolete, or broken electrical or electronic devices. Some electronic scrap components, such as CRTs, contain contaminants such as lead, cadmium, beryllium, mercury, and brominate flame retardants. "Electronic waste" may be defined as all secondary computers, entertainment device electronics, mobile phones, and other items such as television sets and refrigerators, whether sold, donated, or discarded by their original owners. This definition includes used electronics which are destined for reuse, resale, salvage, recycling, or disposal. There is an estimate that the total obsolete computers originating from government offices, business houses, industries and household is of the order of 2 million nos. Manufactures and assemblers in a single calendar year, estimated to produce around 1200 tons of electronic scrap. It should be noted that obsolence rate of personal computers (PC) is one in every two years. At present Bangalore alone generates about 8000 tonnes of computer waste annually and in the absence of proper disposal, they find their way to scrap dealers.

Activists claim that even in developed countries recycling and disposal of e-waste may involve significant risk to workers and communities and great care must be taken to avoid unsafe exposure in recycling operations and leaching of material such as heavy metals from landfills and incinerator ashes.

7.7.1 Recycling

Computer monitors are typically packed into low stacks on wooden pallets for recycling and then shrink-wrapped. Reuse is an option to recycling because it extends the lifespan of a device. Devices still need eventual recycling, but by allowing others to purchase used electronics, recycling can be postponed and value gained from device use.Today the electronic waste recycling business is in all areas of the developed world a large and rapidly consolidating business. Electronic waste processing systems have matured in recent years, following increased regulatory, public, and commercial scrutiny, and a commensurate increase in entrepreneurial interest. Part of this evolution has involved greater diversion of electronic waste from energy-intensive downcycling processes (e.g., conventional recycling), where equipment is reverted to a raw material form. This diversion is achieved through reuse and refurbishing. The environmental and social benefits of reuse include diminished demand for new products and virgin raw materials (with their own environmental issues); larger quantities of pure water and electricity for associated manufacturing; less packaging per unit; availability of technology to wider swaths of society due to greater affordability of products; and diminished use of landfills.

Audiovisual components, televisions, VCRs, stereo equipment, mobile phones, other handheld devices, and computer components contain valuable elements and substances suitable for reclamation, including lead, copper, and gold. One of the major challenges is recycling the printed circuit boards from the electronic wastes. The circuit boards contain such precious metals as gold, silver, platinum, etc. and such base metals as copper, iron, aluminum, etc. Conventional method employed is mechanical shredding and separation but the recycling efficiency is low. Alternative methods such as cryogenic decomposition have been studied for printed circuit board recycling, and some other methods are still under investigation.

7.7.2 Processing Techniques

In developed countries, electronic waste processing usually first involves dismantling the equipment into various parts (metal frames, power supplies, circuit boards, plastics), often by hand. The advantages of this process are the human's ability to recognize and save working and repairable parts, including chips, transistors, RAM, etc. The disadvantage is that the labor is often cheapest in countries with the lowest health and safety standards. In an alternative bulk system, a hopper conveys material for shredding into a sophisticated mechanical separator, with screening and granulating machines to separate constituent metal and plastic fractions, which are sold to smelters or plastics recyclers. Such recycling machinery is enclosed and employs a dust collection system. Most of the emissions are caught by scrubbers and screens. Magnets, eddy currents, and trammel screens are employed to separate glass, plastic, and ferrous and nonferrous metals, which can then be further separated at a smelter. Leaded glass from CRTs is reused in car batteries, ammunition, and lead wheel weights, or sold to foundries as a fluxing agent in processing raw lead ore. Copper, gold, palladium, silver, and tin are valuable metals sold to smelters for recycling. Hazardous smoke and gases are captured, contained, and treated to mitigate environmental threat. These methods allow for safe reclamation of all valuable computer construction materials An ideal electronic waste recycling plant combines dismantling for component recovery with increased cost-effective processing of bulk electronic waste.

Review Questions

1. Enlist the impact of waste on society.
2. Bring out the activities of Individual & NGO's on environmental protection.
3. Explain in detail the Solid waste management techniques.
4. Explain the various types of Radio-active Waste.
5. What are the characteristics of hazardous waste
6. What are the effects of improper municipal solid waste management?
7. State the measures recommended for proper management of the solid waste.
8. Write notes on disposal of radio-active wastes.
9. Explain the methods of disposal of municipal solid waste.
10. What kinds of problems are created due to Urbanisation?
11. Discuss the influence of environmental parameters and pollution on human health.
12. What are the sources of urban and industrial wastes.
13. What are Municipal Solid Wastes(MSW)
14. What is composting.
15. What is incinerators.
16. What is domestic sewage?

CHAPTER 8

Ethics

Ethics and moral values, Ethical situations, objectives of ethics and its study. Preliminary studies regarding Environmental Protection Acts, Environmental Impact Assessment

8.1 Ethics, Values & Morals

Definitions

(a) ***Values*****: Man's** social nature is his fundamental attribute. Human values are conceptions of basic categories of desires. Values are needs/desires. Values are the rules by which we make decisions about right and wrong, should and shouldn't, good and bad. Values are the link that ties together the personal perceptions, judgments, motives and actions These needs are not necessarily self-centered and some of them might be abstract e.g.liberty, egality, conformity, prosperity. Values are more important and primary than facts in forming and understanding all kinds of human purpose. Five human values (Truth, Care, Peace, Duty & justice) are universal, though values are not always held in the sense of being followed, they are every where held in esteem;

(b) ***Morals*****:** Morals refers only to personal behaviour.It refers to any aspect of human action. Morals are simply those customs, the violation of which is regarded in the community as definitely wrong in the word, they are mores. In the stricter sense; the moral code is that of body of rules in which the individual conscience upholds as constituting right or good. The physician who destroys a monstrously deformed baby, for example may violate the community's moral code, but he remains true to his own moral convictions. For most of our daily occasions mores are nearly synonymous with morals. Morals codes vary from person to person, but the mores characterize the group or the community.

(c) ***Ethics*****:** Can be defined as a theory or a system of moral values. Study of right or wrong. Good and evil, obligations & rights, Justice, Social & political deals.. Ethics can be used as synonymous for morally correct. Ethics is activity and area of inquiry. It is activity of understanding moral values or moral ideals and justifying moral judgments as well as the area of study resulting from that activity. The rules or standards governing the conduct of a person or the members of a profession. Ethics involves defining, anlaysing, evaluating and resolving moral problems and developing moral criteria to guide human behaviour.Critical reflection on what one does and why one does it. Refers only to professional behavior.The basic objective of all morals and ethics is to

1. To use the knowledge for the protection of safety and welfare of public
2. To treat all others in a way similar to how you yourself would like to be treated.

8.2 Objectives of Ethics and its Study

Ethics has several distinct although related meanings. They are,

First**:** Ethics is an activity and over of inquiry, understanding moral values, resolving moral issues and justifying moral judgments to guide engineering practice.

Second**:** Moral problems and issues related to engineering.

Third**:** Codes and standards of conduct of various enggs., groups and societies.

Fourth**:** Obligations, rights and ideals to those engaged in engineering discipline.

8.2.1 Personal, Professional Ethics and Engineering Ethics

The primary goal of ethics is to responsibly confront moral issues raised by technological activity and to recognize and resolve moral dilemma.

Personal ethics can be defined as the study of morality issues and decisions confronting while considering the individuals in the society.

Professional Ethics can be defined as the set of standards adopted by the professionals so as to see them acting as a professional. What has to be done and what not has to be done within the profession is a matter choice which is done via professional ethics.

Professional ethics concerns one's conduct of behavior and practice when carrying out engineering work. Such work may include consulting, researching, teaching, manufacturing, servicing and writing. The institutionalization of codes of conduct and codes of practice is common with many professional bodies for their members to observe.

Bodies like Indian Medical Council, Bar Council have brought code of ethics to their respective professional people.Example: Doctors Profession, Law Profession.

Engineering ethics is the study of moral issues and decisions confronting individuals and organizations engaged in engineering / profession. It also includes study of related questions about moral ideals, character, policies and relationships of people and corporation involved in technological activities. One of the major functions of the engineering ethics is that it can serve the function of helping to promote responsible engineering practice.

8.3 Ethical Situations / Moral Problems & Dilemma's

Before getting in to the market a product or project goes through various stages of conception, design and manufacture, followed by testing sales and service. As the engineers carryout their tasks, there will be times when their activities will ultimately lead to a product that in unsafe or less than useful. A product may be intentionally designed for early obsolescence, an inferior material may be substituted under pressure of time or budget or produce ultimately harmful effects may not foresee.

8.3.1 Ethical Problems

The following examples are covered by engineering Ethics.

(i) An engineer finds inferior materials are used for lying road and the informed it to his superiors and stopped the work but his boss viewed the case as a minor infraction in the quailing of road so he ordered to resume the work so that the present would not be delayed, the engineer objected and was threatened and with disciplinary action.

(ii) The plants engineers were aware of contents of pesticide amount in the water used in the soft drink company is beyond the accepted limit. Plant supervisor told to the engineers that it was the local government to identify any problems.

(iii) In building dam Narmada river more than one lakh people has to be evacuated from that particular place where they are been there for a long time.

(iv) Taking office supplies for home use (paper, pencil, pen, etc.)

(v) Using phone for personal use on company time

(vi) Putting personal expenses as business expenses.

(vii) Giving false information

These examples show how Ethical problems arise most often when there is difference of judgment or exceptions as to what constitutes the true state of affairs or a proper course of action. The engineer may be faced with contrary opinions from with in a firm, from the client or from the government.

These cases raise many moral questions:

(i) What does the engineer has to do when there is difference in judgment?

(ii) What does the engineer or supervision's directives be the authoritative guide to an engineer conduct?

8.3.2 Ethical Dilemma' in the Profession

Moral dilemmas are the situations in which two or more moral obligations, duties, rights and ideals comes into conflict with one another. For example, Arun project leader who is having a board meeting in a particular day and the same day his friend is also waiting to meet him, after five years. In this situation both the obligations seem very important for Arun. This situation is called us Moral dilemmas. Moral problems in engineering are twice sort of complexity.

(a) ***Vagueness:*** One's inclarity or confusion over the moral consideration and principle to their moral issues.

For example a medical representative gives a gift to a doctor at the time doctor may have confusion whether to accept the gift or not. If he accepts is it called a bride this is called the vagueness.

(b) ***Conflicting reasons:*** Two or more moral principles applicable to moral issues conflicting each other or one principle itself seems to point out two different directions.

(c) ***Disagreement:*** Reasonable disagreement over for interpretation and applicable of moral principles to the given situation.

8.3.3 Steps in confronting moral dilemma

The problems of vagueness, conflicting reasons and disagreement suggest the need for several steps in approaching dilemmas. They are:

(i) Identifying the moral factors and reasons of dilemma
(ii) Gather all available facts
(iii) Rank the moral consideration in the order of importance
(iv) Consider the alternative courses of action
(v) Consult the colleagues and seeking their suggestions
(vi) A careful reasoned judgment by weighing all moral factors and reasons.
(vii) Question your motives, Practice what you preach, Be your own investigative reporter
(vii) Keep your commitments, Learn to say "No" to things for which you have no time, talent, or sincere interest, Build and maintain your integrity

8.3.4 Ethical use of power

Basis of Power-Guidelines for use

- Treat subordinates fairly
- Maintain credibility
- Be cordial and polite
- Make feasible and reasonable request
- Inform subordinates of rules and penalties
- Warn before punishing
- Punish in private

8.3.5 Positive values to be possessed by an Engineer

- Integrity, honesty
- Truthfulness
- Kind heartedness, humility
- Friendliness
- Faith
- Self respect
- Open mindedness
- Creativity
- Civil sense
- Simplicity
- Forgiveness
- Poise –Equilibrium
- Detachment
- They generate positive thoughts

8.3.6 Ethical Standards

Ethical standards can also be defined as the methods that have set in so as the ethics can be followed. These are also approaches towards doing things. Some of the most important approaches of ethical standards are stated as below-

(i) ***The Utilitarianism Approach*:** This mainly favors in bringing about the greatest amount of good that we can.

(ii) ***The Right Approach*:** This approach best protects and respects the moral rights of those affected.

(iii) ***The Fairness or Justice Approach*:** All should be treated equally.

(iv) ***The Common Good Approach*:** Life in community is good in itself and our actions should continue to that life.

(v) ***The Virtue Approach*:** Actions ought to be consistent with certain ideal virtues make better ethical choices.

8.4 The Code of Ethics

Codes of ethics are concerned with a range of issues, including academic honesty, adherence to confidentiality agreements, data privacy, handling of human subjects, impartiality in data analysis and professional consulting, resolution of conflicts etc, They define ideal behavior of a professional for the purpose of enhancing the public image, establish rules of conduct for policing its own members, and encourage value laden decisions for the public good.

8.4.1 Importance of code of Ethics

Codes of ethics state the moral responsibility of engineering as seen by the profession, and as represented by a profession collective commitment to ethics, codes are of a profession collective commitment to ethics , codes are enormously important for the engineers.

(i) ***Serving and protecting the public:*** Engineering involves both advanced expertise that professional, but not the general public have. In this we have to make a bond in which the trustworthiness and trust are two essential factors

(ii) ***Guidance:*** Codes provide helpful guidance concerning the main obligations of engineers. When they are well written they identify the primary responsibility of the engineers

(iii) ***Inspiration:*** These codes are very often a source of inspiration because they tell us what exactly our responsibilities are. They inspire us more and more to make sure that the safety and health of the public is taken care of. It is indeed our responsibility to do so

(iv) ***Educational and mutual understanding:*** Codes can be used by professional societies and classroom prompt discussion and reflection of moral issues

(v) ***Discipline:*** It also helps in defining what is good and what is bad so that can be taken care of. Thus it in turn results in a strange kind of discipline within the people

(vi) ***Contributing to a profession's image:*** Wherever these codes are followed they help in defining the profession as a positive one

8.4.2 Aims of engineering code of ethics

A code of ethics enables us to

- Set out the ideals and responsibilities of the profession
- Exert a de facto regulatory effect, protecting both clients and professionals
- Improve the profile of the profession
- Motivate and inspire practitioners, by attempting to define their raison d'etre (reason that accounts for)
- Provide guidance on acceptable conduct
- Raise awareness and consciousness of issues
- Improve quality and consistency

8.4.3 Limitations of Code of ethics

Codes of Ethics are concerned with a range of issues, including

(i) Members of professional societies are comparatively less and so they don't feel compelled to abide by their codes.

(ii) Engineering codes often have internal conflicts and they.

8.4.4 Code of Ethics of Professional Engineering Societies

As Engineers we are expected to exhibit the highest standards of honesty and integrity. Engineering has a direct and vital impact on the quality of life for all people. Hence Engineers must perform their duties that requires adherence to the highest principles of ethical conduct. Engineers in the fulfillment of their professional duties shall:

- Hold paramount the safety, health and welfare of the people
- Perform services only in areas of their competence.
- Issue public statement only in an objective and truthful manner.
- Act for each employer or client as faithful agents or trustees.
- Avoid deceptive acts - To receive bribes to show favour
- Conduct themselves honorably, responsibly, ethically and lawfully so as to enhance the honour, reputation and usefulness of the profession.

(I) ***IIEES Professional Codes**:* Thus, as stated in the above the codes of ethics are an essential part of the professional life. Now one such codes of ethics have been laid down by IEEE.

A non-profit organization, IEEE is the world's leading professional association for the advancement of technology.

The IEEE name was originally an acronym for the Institute of Electrical and Electronics Engineers, Inc. Today, the organization's scope of interest has expanded into so many related fields, that it is simply referred to by the letters I-E-E-E (pronounced Eye-triple-E). The Exact codes that have been stated by the IEEE is as below.

We, the members of the IEEE, in recognition of the importance of our technologies in affecting the quality of life throughout the world, and in accepting a personal obligation to our profession, its members and the communities we serve, do hereby commit ourselves to the highest ethical and professional conduct and agree:

(i) to accept responsibility in making decisions consistent with the safety, health and welfare of the public, and to disclose promptly factors that might endanger the public or the environment;

(ii) to avoid real or perceived conflicts of interest whenever possible, and to disclose them to affected parties when they do exist;

(ii) to be honest and realistic in stating claims or estimates based on available data;

(iv) to reject bribery in all its forms;

(v) to improve the understanding of technology, its appropriate application, and potential consequences;

(vi) to maintain and improve our technical competence and to undertake technological tasks for others only if qualified by training or experience, or after full disclosure of pertinent limitations;

(vii) to seek, accept, and offer honest criticism of technical work, to acknowledge and correct errors, and to credit properly the contributions of others;

(viii) to treat fairly all persons regardless of such factors as race, religion, gender, disability, age, or national origin;

(ix) to avoid injuring others, their property, reputation, or employment by false or malicious action;

(x) to assist colleagues and co-workers in their professional development and to support them in following this code of ethics.

(II) ***ACM Professional Code:*** The American Association of Computing Machining Code Of Ethics And Professional Conduct contains many issues that a professional has to face to face so some of the codes are stated as below-

(a) ***Contribute to society and human well-being:*** This principle concerning the quality of life of all people affirms an obligation to protect fundamental human rights and to respect the diversity of all cultures. An essential aim of computing professionals is to minimize negative consequences of computing systems, including threats to health and safety. When designing or implementing systems, computing professionals must attempt to ensure that the products of their efforts will be used in socially responsible ways, will meet social needs, and will avoid harmful effects to health and welfare.

In addition to a safe social environment, human well-being includes a safe natural environment. Therefore, computing professionals who design and develop systems must be alert to, and make others aware of, any potential damage to the local or global environment.

(b) ***Avoid harm to others:*** "Harm" means injury or negative consequences, such as undesirable loss of information, loss of property, property damage, or unwanted environmental impacts. This principle prohibits use of computing technology in ways that result in harm to any of the following: users, the general public, employees, employers. Harmful actions include intentional destruction or modification of files and programs leading to serious loss of resources or unnecessary expenditure of human resources such as the time and effort required to purge systems of "computer viruses."

Well-intended actions, including those that accomplish assigned duties, may lead to harm unexpectedly. In such an event the responsible person or persons are obligated to undo or mitigate the negative consequences as much as possible. One way to avoid unintentional harm is to carefully consider potential impacts on all those affected by decisions made during design and implementation.

To minimize the possibility of indirectly harming others, computing professionals must minimize malfunctions by following generally accepted standards for system design and testing. Furthermore, it is often necessary to assess the social consequences of systems to project the likelihood of any serious harm to others. If system features are misrepresented to users, coworkers, or supervisors, the individual computing professional is responsible for any resulting injury.

In the work environment the computing professional has the additional obligation to report any signs of system dangers that might result in serious personal or social damage. If one's superiors do not act to curtail or mitigate such dangers, it may be necessary to "blow the whistle" to help correct the problem or reduce the risk. However, capricious or misguided reporting of violations can, itself, be harmful. Before reporting violations, all relevant aspects of the incident must be thoroughly assessed. In particular, the assessment of risk and responsibility must be credible. It is suggested that advice be sought from other computing professionals.

(c) ***Be honest and trustworthy:*** Honesty is an essential component of trust. Without trust an organization cannot function effectively. The honest computing professional will not make deliberately false or deceptive claims about a system or system design, but will instead provide full disclosure of all pertinent system limitations and problems.

A computer professional has a duty to be honest about his or her own qualifications, and about any circumstances that might lead to conflicts of interest.

Membership in volunteer organizations such as ACM may at times place individuals in situations where their statements or actions could be interpreted as carrying the "weight" of a larger group of professionals. An ACM member will exercise care to not misrepresent ACM or positions and policies of ACM or any ACM units.

(d) ***Be fair and take action not to discriminate:*** The values of equality, tolerance, respect for others, and the principles of equal justice govern this imperative. Discrimination on the basis of race, sex, religion, age, disability, national origin, or other such factors is an explicit violation of ACM policy and will not be tolerated.

Inequities between different groups of people may result from the use or misuse of information and technology. In a fair society, all individuals would have equal opportunity to participate in, or benefit from, the use of computer resources regardless of race, sex, religion, age, disability, national origin or other such similar factors. However, these ideals do not justify unauthorized use of computer resources nor do they provide an adequate basis for violation of any other ethical imperatives of this code.

(e) ***Honor property rights including copyrights and patent:*** Violation of copyrights, patents, trade secrets and the terms of license agreements is prohibited by law in most circumstances. Even when software is not so protected, such violations are contrary to professional behavior. Copies of software should be made only with proper authorization. Unauthorized duplication of materials must not be condoned.

(f) ***Give proper credit for intellectual property:*** Computing professionals are obligated to protect the integrity of intellectual property. Specifically, one must not take credit for other's ideas or work, even in cases where the work has not been explicitly protected by copyright, patent, etc.

(g) ***Respect the privacy of others:*** Computing and communication technology enables the collection and exchange of personal information on a scale unprecedented in the history of civilization. Thus there is increased potential for violating the privacy of individuals and groups. It is the responsibility of professionals to maintain the privacy and integrity of data describing individuals. This includes taking precautions to ensure the accuracy of data, as well as protecting it from unauthorized access or accidental disclosure to inappropriate individuals. Furthermore, procedures must be established to allow individuals to review their records and correct inaccuracies.

This imperative implies that only the necessary amount of personal information be collected in a system, that retention and disposal periods for that information be clearly defined and enforced, and that personal information gathered for a specific purpose not be used for other purposes without consent of the individual(s). These principles apply to electronic communications, including electronic mail, and prohibit procedures that capture or monitor electronic user data, including messages, without the permission of users or bona fide authorization related to system operation and maintenance. User data observed during the normal duties of system operation and maintenance must be treated with strictest confidentiality, except in cases where it is evidence for the violation of law, organizational regulations, or this Code. In these cases, the nature or contents of that information must be disclosed only to proper authorities.

(h) ***Honor confidentiality:*** The principle of honesty extends to issues of confidentiality of information whenever one has made an explicit promise to honor confidentiality or, implicitly, when private information not directly related to the performance of one's duties becomes available. The ethical concern is to respect all obligations of confidentiality to employers, clients, and users unless discharged from such obligations by requirements of the law or other principles of this Code.

8.5 Environmental Ethics

Definition

Environmental ethics refers to the issues, principles and guidelines relating to human interactions with their environment.

It is a branch of applied ethics associated with the restoration of natural environment in a balanced state by not harming the human society through vast industrialization is called environmental ethics. Nature is a valuable resource for recreation as well as economic development. The greater efforts to protect our environment are required is not controversial among people who have given the matter serious thought, as we are all sharing common environment.

(a) Sources of Environmental ethics

Anthropocentric ethics: It is the belief that ethics and law are inadequate to deal with the vast array of environmental problem. It prefers cost-benefit analysis of decision that affects environment, by taking adequate account of the harm to human interests that environmental damage can cause

Bio-centric ethics: It is the belief that nature, wildness area and non-human life have their own inherent value and therefore deserves moral considerations. All human decisions that affects environment should incorporate the respect for nature.

(b) Major Environmental problems/issues

(i) Air pollution & water pollution
(ii) Ozone layer depletion, Global warming & Acid rain
(iii) Deforestation activities
(iv) Degradation of soil fertility
(v) Population growth & Urbanization
(vi) Land degradation and water scarcity.
(vii) Loss of biodiversity

(c) Technology cost of Environmental degradation: Engineering projects & experiments need careful monitoring by assessing the technology that includes social and environmental effects of the technology. The technology of production, from agriculture to manufacture of plastics, the efficiency is figured out through the cost-benefit analysis and the approximated price that involves direct costs. The true cost would have to include numerous indirect factors such as effects of pollution, depletion of energy and the raw materials and the social costs.

(d) **Solutions to Environmental Problems:** The environment can be protected due to the following activities.

(i) Reduce the waste of matter and energy resources

(ii) Recycle and reuse as many of our waste products and resources as possible

(iii) Over-exploitation of natural resources must be reduced

(iv) Soil degradation must be minimized

(v) Sustanible development is essential, conservation of natural resources, harvesting non-conventional energy and waste management

(vi) Biodiversity of the earth must be protected

(vi) Reduce population and increase the economic growth of our country.

(e) **Ethical Guidelines**

(i) You should love and honor the earth since it has blessed you, with life and governs your survivals

(ii) You should keep each day sacred to earth and celebrate the turning of its seasons

(iii) You should not harm other living things and have no right to drive them to extinction

(iv) You should be grateful to the plants and animals, which nourish you by giving you food.

(v) You should not waste your resources on destructive weapons

(vi) You should not steal from future generation their right to live in a clean and safe planet by polluting it.

(vii) You should consume the materials goods in moderate amounts so that all may share the earth's precious treasure of resources.

8.6 Environmental Protection Acts

Constitution of India has a number of provisions demarcating the responsibility of the state and central government towards 'Environmental protection'. The state's responsibility has been laid down under article 48-A which reads as follows, "the state shall endeavor to protect and improve the environment and safeguard the forests and wildlife of the country". Environmental protection has been made a fundamental duty of every citizen of this country under article 51-A (g) which reads as "it shall be the duty of every citizen of India to protect and improve the natural environment including forest, lakes, rivers and wildlife and to have compassion for living creatures". Article 21 reads as, "no person shall be deprived of his life or personal liberty except according to procedure established by law".

***Definition of environment under Indian law*:** According to section 2(a) of Environment Protection Act (1986) , Environment includes, (1) water, air and land, (2) the interrelationship which exists among and between, (a) water, air and land (b) human beings, other living creatures, plants, microorganisms and property.

Various statues / legislation or enacted in India exclusively for environment protection are,

(a) The Water (Prevention and Control of Pollution) Act, 1974.
(b) The Air (Prevention and Control of Pollution) Act, 1981.
(c) The Environmental Protection Act, 1986.
(d) The Forest Conservation Act, 1980.
(e) The Wildlife Protection Act, 1972.
(f) The Public Liability Insurance Act, 1991 etc.

8.6.1 Air (prevention and control of pollution) act, 1981

'Air Pollution' means the presence in the atmosphere of any air pollutant. Air pollution means any solid , liquid or gaseous substances (including noise) present in the atmosphere in such concentration as may be or tend to be injurious to human beings or other living creatures or plants or property or environment.

The objective of the Act is to provide for the prevention, control and abatement of air pollution for the establishment with a view to carry out the aforesaid purpose of boards for conferring on and assigning to such boards powers and functions relating there and for matters connected therewith.

***Functions of Central Board*:** The main functions of central board as specified in section 16 of the act shall be to improve the quality of air and to prevent, control or abate air pollution in the country.

(a) Advice to central government on any matter related to air quality.
(b) To execute wide awareness programme.
(c) Co-ordinate with state boards and resolve disputes among them.

(d) To provide technical assistance and guide to state boards.
(c) Sponsor research and investigating regarding problem of air pollution.
(e) Collect technical and statistical data to prepare manuals, code, guide related to air.
(f) To lay down standards for quality of air.

Importance of Various Sections of Air Act:

Section 19 – Declaration of air pollution control area

Section 10 – Lays down the standards for emission of air pollutants from automobiles

Penalty for contravention of Certain Provision of the Act: Whoever contravenes any of the provisions of this act or any order or decision issued there under for which no penalty has been elsewhere provided in this act shall be punishable with imprisonment for a tern which may extend to 3 months or with a fine extend to Rs. 10,000/- or with both.

Both companies and government departments are also prosecuted under this Act. No court shall take cognizance of any offence except on a complaint made by any person who has given notice of not less than 60 days, in the manner prescribed of the alleged offence and his intention to make a complaint to the board.

8.6.2 Water pollution act

The objective of the Water Prevention and Control of Pollution Act was to provide for the prevention and control of water pollution and maintenance or restoring of wholesomeness of water for the establishment with a view to carry out the purpose aforesaid, of boards for the prevention and function relating thereto and for matters connected therewith.

Functions of Central Board:

(a) Promote cleanliness of streams and wells in different areas of the state.
(b) Advice the central government on any matter concerning the prevention and control of water pollution
(c) Co-ordinate the activities of state boards and resolve disputes among them.

(c) Provide technical assistance and guidance to the state board, carryout and sponsor investigations and research relating to problems of water pollution.

(d) Organize through mass media, a comprehensive programme regarding the prevention and control of water pollution.

(e) Collect, compile and publish technical and statistical data relating to water pollution and the measure devised for its effective prevention and control and prepare manuals, codes regarding the treatment and disposal of sewage and trade effluents.

(f) Establish and recognize a laboratory to enable the board to perform its functions under this section effectively, including the analysis of sample of water from any stream or well of sample of any sewage or trade effluents.

Functions of State Board:

(a) To plan a comprehensive programme for the prevention, control or abatement of pollution of stream and well in the state and to secure the execution there of.

(b) To advise the state government on any matter concerning the prevention, control or abatement of water pollution

(c) To collect and disseminate information relating to water pollution, prevention, control or abatement of water pollution

(d) To encourage, conduct and participate the investigations and research relating to problems of water pollution

(e) To collaborate with central board in organizing the training of persons engaged in programmes relating to water pollution, prevention, abatement and treatment.

(f) To inspect effluent treatment plants trade waste and domestic waste.

(g) To laydown, modify standard for trade and domestic wastes.

(h) To evolve economical and reliable methods of treatments, utilization of treated effluent for agriculture and disposal into land.

(i) To lay down standards for treatment of sewage and trade effluents to be discharged into stream during dry weather flow.

(j) To advise state government with respect to the location of any industry the carrying on which is likely to pollute a stream or well.

Importance of Section 24 of Water Act, 1974: No person should knowingly cause or permit any poisonous, noxious or polluting matter determined in accordance with such standards as may be laid down by the state board to enter into any stream or well or sewer or on land.

However, a person shall not be party of an offence under subsection (1), by any reason only of having done or could to be done by any of the following acts namely;

(a) Constructing bridge, weir, dam, sluice, dock, pier, drain, or sewage or other permanent works which he has a right to construct, improve or maintain.
(b) Depositing any material on the bank or in the bed of any stream for the purpose of reclaiming land or for supporting repairing or protecting the bank or bed of such stream provided such material are not capable of polluting such streams.
(c) Polluting into any stream by any sand or gravel or other natural deposit which has flowed from or been deposited by the current of such stream.

Whoever contravention of provisions of section (24) shall be punishable with imprisonment up to six years and with fine. Even the municipality corporation, companies, government departments also be prosecuted under water act. Varieties of powers are given to the central / state boards to make application to courts for restrains apprehended pollution of water in streams and wells

8.6.3 Wild life protection act

The rapid decline in India's wild life (animals and birds). One of the richest and most varied in the World has been a cause of concerns. It is really lamentable that some animals and birds species have already become extinct: some more species are under grave threats for their existence. The Wild Birds and Animals Protection Act 1912 has almost become outmoded. The emphasis of earlier act was on control of hunting. Other important aspects like taxidermy, and trade in wildlife were not attended to. Need was felt for the enactment of a comprehensive legislation and the wildlife protection Act 1972 was brought into force.

The salient features of the Act: The Act contains 66 sections in seven chapters and six schedules. The definitions of the terms 'Animals', 'Animal article', 'Captive Animal', 'Hunting', 'Wild Animal', 'Wild Life', etc. are exhaustive and aimed to provide a broader scope of operation for the law.

The Act provides for appointment of authorities like Director of wild life preservation. Asst. Director at the center, chief wild life warden and other officers in the states. Provision is there for the constitution of wildlife Advisory board in each state.

The Act prohibits hunting of wild animals (specified in the schedule), protection for the specified plants, establishment of sanctuaries, national parks, etc. the schedules in the act are important.

Schedule I prohibits harming endangered species throughout India. It is important to remember Art 51 (A) (g) of the Indian constitution which prescribes that it is the duty every citizen 'to protect and improve the natural environment including forests, lakes, rivers, and wildlife and to have compassion for all living species'.

In 1982 there was amendment to the act which introduced provisions permitting the capture and transport or wild animals for scientific management of the animal populations.

8.6.4 Forests conservation act, 1980

'Non Forest Purpose' means the breaking up of cleaning of any forest, land or portion thereof for the cultivation of tea, coffee, spices, rubber, palms, oil bearing plants, horticultural crops, medicinal plants or plantation crops.

It is well known that breaking up the soil or clearing of the forest land affects seriously reforestation or regeneration of forests and therefore, such breaking up of soil can only be permitted after advantages and disadvantages to the economy of the country. Environmental conditions, ecological imbalance that is likely to occur, its effect on the flora and the fauna in the area, etc., it was therefore thought that the entire control of the forest areas should vest in the central government. With that end in view, Section 2 provided that prior approval of the central government should be obtained before permitting the use of the forest land for non-forest purpose.

Current Requirements that should be met before declaring an area into a Wild Life Sanctuary / National Park under Forest Act.

(a) The state government may by notification in the office declare the provision of their chapter applicable to any forest land or waste land which is not included in a reserve forest, but which is the property of government.

(b) The forest land and waste land included in any such notification shall be called a 'protected forest'.

(c) No such notification shall be made unless the nature and extent of the rights of government and of private persons in or over the forest land or waste land comprised therein have been inquired into and recorded at a surveyor settlement, or in such other manner as the state government thinks sufficient.

Section 35 *– Protection of Forest for Special Purpose:*

(i) The state government may, by notification in the official Gazette, regulate or prohibit in any forest or waste land.

(ii) The state government may, for any, such purpose, construct on its own expense, in or upon any forest or wasteland, such work on it thinks fit.

(iii) No notification shall be made under subsection (1) nor shall any work begun under subsection (2) until after the issue of notice to the owner of such forest or land calling on him to show cause , within a reasonable period to be specified in such notice , why such notification shall not be made or work constructed , as the case may be and until such objection , if any evidence he may produce in support of the same , have been heard by an officer duly appointed for that purpose and have been considered by the state government.

8.6.5 Environmental protection act, 1986 (EPA)

Terms like 'environment', environmental pollutants, environmental pollution and hazardous substance defined under EPA 1986.

(a) 'Environment' includes water, air and land and the interrelationship which among and between them and human beings, other living creatures, plants, micro- organisms and property.

(b) 'Environmental pollutant' means any solid, liquid or gaseous substances present in such concentration as may be or tend to be injurious to environment.

(c) 'Environment pollution' means the presence in the environment of any environmental pollutants.

(d) 'Hazardous Substance' means any substance or preparation which by reason of its chemical or physico-chemical properties or handling is liable to cause harm to human beings, other living creatures, plants, micro-organisms, property of the environment.

General Powers of the central government under EPA**:** Subject to the provisions of the Act, the central government shall have the power to take all such measures as it seems necessary or expedient for the purpose of protecting and improving the quality of the environmental pollution.

In particular and without prejudice to the generality of the provisions of sub-section (1) such measures may include measures with respect to all or any the following matters.

(a) Co-ordination of actions by the state government officers.
(b) Planning and execution of nationwide programme on 'Environment Pollution'.
(c) Laying down standards for emissions or discharge of environmental pollutants from various sources whatsoever.
(d) Laying down procedures and safeguards for the prevention of accidents which may cause environmental pollution.
(e) Laying down procedures to safeguard hazardous substances.
(f) Examination of such manufacturing process, materials and substances as are likely to cause environmental pollution.
(g) Carrying out and sponsoring investigation and research.
(h) Inspection of the premises, plants, equipment, machinery, manufacturing or other processes, material or substances.
(i) Establishment or recognition of Environmental laboratories and institutions to carry out functions entrusted to them.
(j) Preparation of manuals, codes, guides, etc.

Section 4 – Appointment of officers and their powers and functions

Section 5 – Power to give directions

Section 6 – Rules to regulate Environmental pollution

Under EPA pollution of land and soil is also covered. Penalties for violation under EPA are also listed. Companies and government may also be prosecuted under EPA.

8.7 Environmental Impact Assessment

Human activities create environmental impacts. The effects of these activities can be felt during their construction, and operation. It becomes difficult to mitigate or avoid the ill effects after establishing the project. Therefore the impacts that may arise later have to be visualized before hand so that the developmental activities are harmonized with the environment. The exercise of visualizing or assessing the effects of a project on the environment before taking it up is called as 'Environmental Impact Assessment (EIA)'. EIA makes it possible to integrate the environmental aspects into the developmental activities during initiation of the project. It prevents the environmental and economic liabilities that may arise in future. A proposed project can be shelved in the beginning itself if it is found to be detrimental to the environment.

EIA is conducted step by step in a systematic way.

Steps in an EIA study

***Step*: 1** Description of the project and the site of construction. Water and raw material requirement is estimated. Industrial processes, production etc are described.

***Step*: 2** Alternative sites for the project are evaluated for consideration

***Step*: 3** Base line data collection – It describes the existing environmental status of the study area which is the area covered in a certain radius with proposed project / industry as the centre.

In the baseline study data on the following aspects are collected:

- Land and land use pattern.
- Existing water resources - quantity and quality wise
- Air quality
  (a) Meteorology and climate data such as temperature, wind speed and direction, rainfall, humidity etc.,
- Soil quality
- Seismological characteristics
- Noise, and traffic
- Biological environment
  (a) Plant species
  (b) animal species
  (c) endangered species
- Agriculture potential
- Historical sites and monuments
- Tourist spots
- religious centres
- Wild life sanctuaries
- Schools, hospitals etc.,
- Demography, cultural and socio economic environment
- Any other environmentally significant parameter

The possible impacts of the proposed project on the existing environmental setting are assessed by superimposing the effects of the project on the existing environment. If the impacts are not acceptable, corrective measures are incorporated into the proposed project and then correlated with the existing environmental set up. If significant negative effects are not observed, the project can be permitted to be taken up. In case, even after taking protective measures the environment is going to be affected, permission will not be given to establish the proposed project. In the EIA exercise public are also allowed to participate and express their opinion. Based on the outcome of the EIA studies a status report called 'Environmental Impact Statement (EIS)' is prepared which serves as a guideline for establishing environmentally sustainable activity.

In India Ministry of Environment and Forest (MOEF) guides and controls the EIA process through the state pollution control boards.

Review Questions

1. Differentiate ethics, morals and values
2. What is the need to focus on professional ethics?
3. Define Personal and Professional ethics
4. What are the steps in confronting moral dilemmas?
5. Discuss the scope and aims of engineering ethics
6. Discuss any two case studies on professional disagreement an engineer may encounter and discuss how you would act in that situation
7. Where and how do moral problems arise in engineering?
8. Give the steps in confronting moral dilemmas
9. Explain the skills needed to handle problems about moral issues in engineering ethics.
10. What are the responsibilities of engineers to society?
11. Discuss the role played by codes of ethics set by professional societies.
12. How do you internalize the costs of environmental degradation?
13. Explain environmental ethics with a case study.
14. Name the laws that have been framed for environmental protection and mention the objectives and features of each act.
15. Explain the steps involved in EIA Study.

www.ingramcontent.com/pod-product-compliance
Lightning Source LLC
LaVergne TN
LVHW021812240826
846425LV00002B/17

* 9 7 8 9 3 8 5 4 3 3 1 5 3 *